THE ILLUSTRATED FIVE KINGDOMS

Dedicated to the memory of Professor HEINZ A. LOWENSTAM (1912–1993), beloved teacher, natural historian, geologist, marine biologist and founder of the field of biomineralization

The Illustrated Five Kingdoms

A Guide to the Diversity of Life on Earth

Lynn Margulis
University of Massachusetts, Amherst

Karlene V. Schwartz
University of Massachusetts, Boston

Michael Dolan
University of Massachusetts, Amherst

Illustrated by
Kathryn Delisle and Christie Lyons

HarperCollinsCollegePublishers

Cover illustration
A = African tree pangolin, *Manis tricuspis*;
B = American elm, *Ulmus americana*;
C = fly agaric, *Amanita muscaria*; **D** = velvet worm, *Peripatus*;
E = cyanobacterium, *Anabaena*; **F** = colon bacterium, *Escherichia coli*; **G** = anaerobic phototrophic bacterium, *Chromatium*; **H** = bladder kelp, *Nereocystis*;
J = chambered nautilus, *Nautilus pompilius*.

Christie Lyons

Acquisitions Editor: Susan McLaughlin
Editor/Designer: Stephanie Hiebert
Editorial Assistant: Landi Stone
Production Manager: Robert Cooper
Printer and Binder: Malloy Lithographing, Inc.
Cover Printer: Lehigh Press Lithographers

The Illustrated Five Kingdoms: A Guide to the Diversity of Life on Earth

by Lynn Margulis, Karlene Schwartz, and Michael Dolan

Library of Congress Cataloging-in-Publication Data
Margulis, Lynn, 1938–
    The illustrated five kingdoms: a guide to the diversity of life on earth / Lynn Margulis, Karlene V. Schwartz, Michael Dolan; illustrated by Kathryn Delisle and Christie Lyons.
        p. cm.
    Includes bibliography references and index.
    ISBN: 0-06-500843-X
    I. Schwartz, Karlene V., 1936– . II. Dolan, Michael. III. Title.
    QH313.M37 1994
    574—dc20
                                                                        93-6138
                                                                        CIP

    94 95 96 97 9 8 7 6 5 4 3 2

Contents

Note to the Reader

As we burn rain forests for farm space, ranchland, and hardwood, as we drain ponds and level hills to accommodate housing projects and shopping malls, we encroach on the lives and living space of the other organisms with whom we share this planet. We are clearing out our planetmates faster than we are learning about them. The relentless increase of human beings at the cost of other organisms has inspired many eloquent scientists and concerned citizens to put the study of biodiversity—different kinds of life—high on the social agenda. The small guide you hold in your hands outlines the basics of biodiversity. To introduce the reader to the diminishing life on Earth, we have selected at least one member of each of the largest groupings, or phyla, of the five kingdoms and depicted it with the other members of its living community.

Designed for teachers and students, as well as the interested public, the book's spiral binding makes production of multiple copies for coloring or labeling easy. Because descriptions are subject to differing opinions and change as new information becomes available, no names or labels appear on the full-size illustrations. A reduced, labeled copy of each drawing accompanies the text that describes the organisms in their habitat. The genus (which comes first but is more like our family name) and species (second, but more like our personal name) of most organisms, as well as explanatory information, are provided. A complete list of species, genera, and phyla illustrated in this book appears in the Appendix, beginning on page 165. Members of some of the most abundant and familiar phyla (such as the molluscs, chordates, and angiosperms) are represented more than once.

Introduction

Biodiversity: All of Life

Although full of astounding beings, much of life on Earth remains a mystery. This guide introduces many unfamiliar organisms and describes many familiar ones. Through illustrations that show examples from each of the phyla, the full range of living beings is discussed. This book is a primer in biodiversity.

Unlike other books on biodiversity, this work places each organism in its habitat, i.e., its natural environmental context. From photosynthetic bacteria to amebas and slime molds, from horsetails to tube worms and velvet worms, each being is depicted in its own surroundings, which includes other members of its community. In the context of natural history, ecological, taxonomic, and structural information are presented together.

We tend to take for granted the live organisms that do not belong to our species, even though they include all of those upon whom we depend for life. Our food, shelter, and pharmaceuticals come from other living beings. Furthermore, coal and petroleum products, including gasoline, plastics, and motor oil, all come from traces of bygone that were left on Earth's surface when great algal crusts or giant forests died and were buried before they could decay entirely. Although no one knows for sure, it is estimated that there are roughly thirty million species of organisms in total, of which more than 99.9% are extinct. Most present-day species are insects or other arthropods. As many as 100,000 species of fungi and 250,000 species of protoctists are estimated to exist. There may be between 250,000 and 500,000 species of plants, but since most are in the tropics where they are least studied, no one knows for certain. As for bacteria, the biological literature lists some 10,000 different types, but most are probably still unknown.

Autopoiesis, or What Is Life?

All living beings share certain characteristics: They are composed of cells bounded by membranes composed of proteins and lipids (fatty substances). Inside the cells are biochemicals, including several thousand genes (stretches of DNA) and proteins (stretches of linked amino acids). DNA is made of small molecules hooked together in a sequence that determines the order of the amino acids in proteins. These intracellular materials actually produce the membranes, proteins, and nucleic acids that make up the cells. Thus, cells are self-making and self-maintaining. External sources of energy and nutrients provide the raw

1

materials for the components of cells to repair their parts and make more of themselves. This process of self-maintenance and growth, a property of all live beings, is called autopoiesis.

The word used to refer to the sum of the chemical transformations that underlie the autopoietic processes is metabolism. Dead bodies, which may at first have the same composition as their live counterparts, do not metabolize; by contrast, all live organisms metabolize. Metabolism involves the exchange of gases, chemical compounds, and other materials and energy with the surroundings. Eventually cells that have grown large may reproduce. The reproduction of cells is based on the replication, or copying, of their nucleic acid molecules: DNA and RNA. Replication is a property of molecules (DNA can copy itself in a test tube); reproduction is a property of life. Although DNA and RNA molecules can replicate, nothing less complex than a cell or organism composed of cells can reproduce.

Viruses

Viruses, which are much smaller and much less complex than cells, are not autopoietic. If fed, watered, and supplied with chemical or light energy, they still cannot metabolize. Viruses can replicate but cannot reproduce on their own. Even replication of viruses requires that they enter actively metabolizing cells. All members of the five kingdoms are composed either of prokaryotic (nonnucleated) cells (i.e., bacteria) or eukaryotic (nucleated) cells (i.e., protoctists, fungi, animals, and plants). Because they are not cellular entities, viruses do not belong to any of the five kingdoms of organisms. Alternatively, viruses may be thought of as inanimate parts of the cells that they infect and hence can be classified with these cells. Some viruses are depicted in Figure 1.

How Many Species Are There?

We estimate that there are more than thirty million known species, of which most are arthropods (the animal phylum to which all insects belong) or nematodes. Many thousands more species await discovery. Probably 99% of living microorganisms still are not known to science.

Cells and Movement: Flagella and Undulipodia

The parts of cells move. Whether the cell is prokaryotic (bacterial) or eukaryotic, its components bounce around by random jiggling called Brownian motion. In addition, bacterial cells may glide slowly if attached to a surface (by an unknown mechanism) or may swim using their rotary motor flagella (Figure 2).

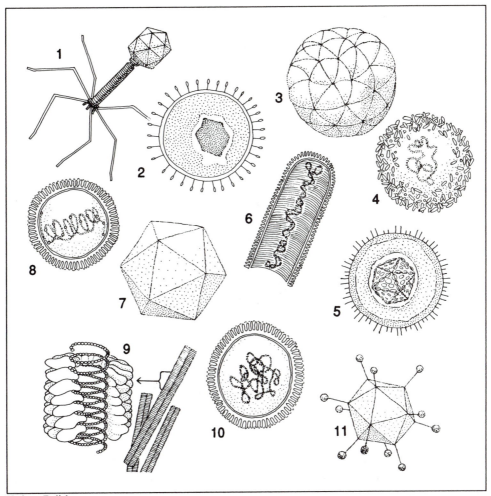

Kathryn Delisle

Figure 1
Viruses are particles of DNA or RNA enclosed in a protein coat. Because they are not made of cells, viruses are not considered to be living organisms. Although they reproduce, they can do so only by entering a cell and using its living machinery to replicate themselves. Outside of the cell a virus cannot reproduce, feed, or grow. The viruses pictured here are (1) bacteriophage, (2) human immunodeficiency virus (HIV), (3) polio virus, (4) pox virus, (5) herpes virus, (6) rhabdovirus, (7) picorna virus, (8) orthomyxovirus, (9) tobacco mosaic virus (TMV), in cross section (left) and whole (right), (10) paramyxovirus, and (11) adenotype 2 virus.

Bacterial cells divide by direct division, a process called binary fission. Although eukaryotic cells also divide in two, they do not do so directly: They divide by mitosis. During many mitotic cell divisions, at the ends (poles) of the cells are two or four dotlike bodies that, magnified, show a characteristic structure of arrayed microtubules: nine triplet tubules in a cylindrical pattern. The dotlike bodies are centrioles. The centriole structure of nine groups of three microtubules each and no central microtubules is called the [9(3)+0] array. Eukaryotic cells swim using wavy structures, cilia, or other kinds of undulipodia. These

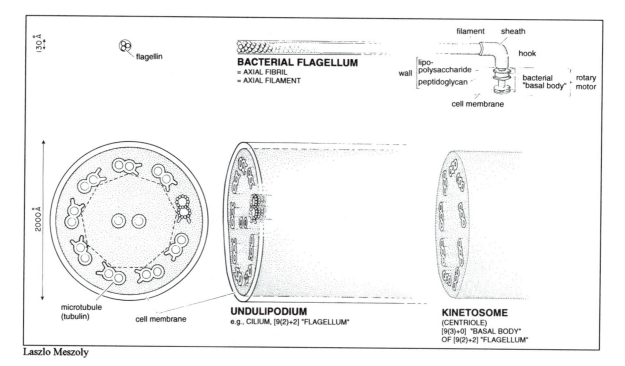

Laszlo Meszoly

Figure 2
Although both move the organism through its environment, prokaryotic flagella (top) and eukaryotic undulipodia (bottom) differ fundamentally in structure.

undulipodia—whether oviduct cilia, sperm tails, or amebomastigote undulipodia—show the same pattern in cross section under an electron microscope. The underlying kinetosome is identical to the centriole, but the shaft, with its nine groups of two fused microtubules each, surrounding two separate central microtubules, has a slightly different array (see Figure 2).

Undulipodial shafts have [9(2)+2] cross sections made of longitudinally aligned microtubules. All undulipodia are underlain by kinetosomes (Figure 3), basal structures from which they grow. Undulipodia are modified to make a fantastic variety of functional structures, including the sensory hairs of lobster antennules and the cilia of the gills of marine animals. The term used to describe mitotic centrioles and cilia basal bodies is centriole-kinetosome. Centriole-kinetosomes, undulipodia, mitosis, and [9(2)+2] arrays are entirely absent in prokaryotes—i.e., they do not occur in members of the kingdom Monera. Although microtubules and mitosis are present in fungi, centriole-kinetosomes and undulipodia are always absent.

Because of the confusion surrounding the names of these structures, we define them as follows:

Flagella (Greek: "whip"), **s. flagellum.** Solid bacterial organelles of motility composed of flagellin, intrinsically nonmotile. Rotary locomotion is generated at the points of insertion of organelle into the cell. Flagella

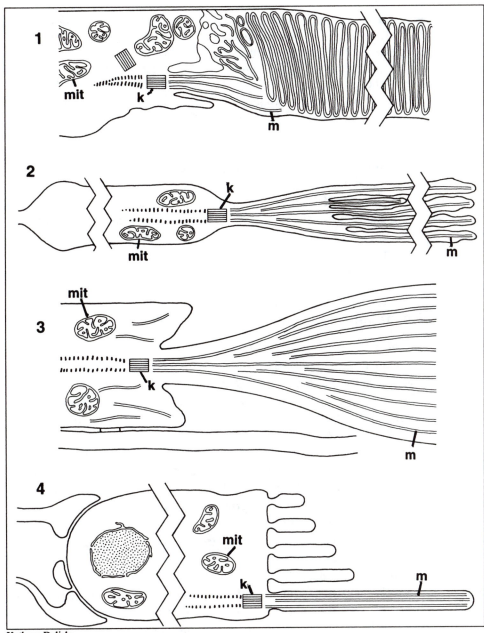

Kathryn Delisle

Figure 3
Four general types of animal sensory cells based on cilia (a type of undulipodium). All contain modified undulipodia with microtubules. (1) Rod cell of the eye (vision). (2) Olfactory cell (smell). (3) Mechanoreceptor cell (touch). (4) Inner-ear cell (hearing). **k** = kinetosome; **m** = microtubule; **mit** = mitochondrion; zig-zags represent omissions in long structures.

are extracellular in that they always extend externally beyond the plasma membrane of the prokaryotic cell. Diameter: 15 to 30 nanometers. Includes the axial filaments or axial fibrils of spirochetes. In these latter microbes the flagella are situated in the periplasm, the space between the outer and the inner (plasma) lipoprotein membranes of the Gram-negative cell wall.

Undulipodia (from Latin: *undula,* "a little wave," and Greek: *podos,* "foot": "little waving feet"), **s. undulipodium.** Cilia and [9(2)+2] flagella: Long slender tubulin-containing intracellular organelles of motility of eukaryotes, intrinsically motile throughout their length, capable of movement when severed from the cell. Diameter: 0.25 micrometer (250 nanometers). The axoneme is the [9(2)+2] shaft of the undulipodium; i.e., it is the undulipodium lacking its surrounding membrane.

Many biologists retain "flagellum" and "basal body" for the ninefold symmetrical microtubular motility organelles of eukaryotes. We suspect that they are reluctant to change because they do not generally confront bacteriologists. In the past the term flagellum has referred both to the undulipodium/cilium found in animals, plant sperm, and many protoctists, as well as to the rotating bacterial organelle. This ambiguous use of the term has led to confusion, which, by using different terms, we avoid in this book.

Modes of Nutrition: Autotrophy and Heterotrophy

The existence of photosynthetic animals such as *Convoluta roscoffensis* and heterotrophic plants such as Indian pipes (*Monotropa*) makes clear that nutritional mode alone cannot be used to define the highest taxa. Indeed, microbes display a remarkable diversity of nutritional modes (Table 1). With the exceptions of photoheterotrophy and chemotrophy, every nutritional mode is represented among the protoctists.

Photoheterotrophs use organic compounds as food sources while simultaneously employing visible light to generate ATP directly, whereas chemoautotrophs can use exclusively carbon dioxide and other inorganic compounds as sources of carbon and energy. In strict chemoautotrophy (e.g., methanogenesis, methylotrophy, ammonia oxidation, sulfide oxidation, and the like), neither the source of carbon nor the source of energy is from carbon–hydrogen (organic) compounds. These two metabolic modes— photoheterotrophy and chemoautotrophy—are found only in bacteria. Thus, although they display a greater range of nutritional types than do plants, fungi, or animals, protoctists are far more limited in energy- and nutrient-gathering capability than are bacteria. Bacteria are by far the most nutritionally diverse organisms; photosynthesis and motile heterotrophy evolved in them.

All five kingdoms include organisms that exhibit biomineralization—the production of minerals by live cells and the incorporation of those minerals into their bodies or protective coverings (Table 2).

MODE/EXAMPLE	ENERGY SOURCE	CARBON SOURCE
Autotrophy		
Plants, algae, and cyanobacteria	Light	Atmospheric CO_2
Sulfide-, methane-, and ammonia-oxidizing bacteria	Inorganic compounds (H_2S, CH_4, and NH_3)	Atmospheric CO_2
Heterotrophy		
Fungi, protoctists (ciliates, mastigotes, slime nets, etc.) and animals (molluscs, hydras, fish)	Organic compounds (C, H, N, O)	Organic compounds (C, H, N, O)

Table 1: Modes of Nutrition
Organisms have evolved various ways of obtaining energy and carbon—the two requirements for growth and reproduction.

Reproduction or Sex

At least some organisms from each of the five kingdoms reproduce directly—i.e., the offspring have only one parent. Generally referred to as asexual reproduction, this process is better called simply reproduction. Reproduction is defined as any process in which the number of living beings increase. When we are wounded, our cells reproduce by mitotic division, and at least some members of most animal groups make new multicellular individuals directly from one parent (e.g., sponge gemmae, hydra buds, and coral strobili). All members of the moneran, protoctist, and fungal kingdoms reproduce by propagules that are produced by a single parent (i.e., asexuality is the norm). Animals, however, which develop from embryos formed by a sexual act (fertilization of an egg by sperm), and plants, which grow after an egg nucleus is fertilized by a sperm or pollen nucleus, are sexual.

Sexuality is entirely different from reproduction, in principle, even though the two are intertwined in the most familiar organisms: mammals and flowering plants. In biology, sex or sexuality is the formation of a new organism from more than a single genetic source, usually from the genes of two parents. Therefore, an individual who has developed from a sexual event has two biological parents rather than one. By definition, then, sex begins with two sources of DNA (the genes of the two parents) and results in the formation of a new individual (called a recombinant individual) with genes from both sources.

Bacteria engage in a peculiar form of sex in which they inject their genes into each other. Any bacterium that has its own genes and receives other genes from a fellow bacterium is, by definition, the product of a sexual event, even though the cells do not fuse. (Only the DNA molecules pass from cell to cell.) The contact of bacterial cells followed by the one-way injection of genes from donor to recipient is called conjugation, or bacterial mating. Cell fusion is limited

MINERAL	MONERA	PROTOC-TISTA	FUNGI	ANIMALIA	PLANTAE
Calcium					
Calcium carbonate ($CaCO_3$: aragonite, calcite, vaterite)	Extracellular precipitates Sheath precipitates	Ameba tests Foraminiferan tests	Extracellular precipitates	Corals Mollusc shells Echinoderm skeletons Calcareous sponges Some kidney stones	Precipitates
Calcium phosphate ($CaPO_4$)			Extracellular precipitates	Brachiopod "lamp shells" Vertebrate teeth and bones Some kidney stones	
Calcium oxalate (CaC_2O_4)				Most kidney stones	*Dieffenbachia*
Silicon					
Silica (SiO_2)	Precipitates	Diatom tests Radiolarian tests Silicomastigote scales		Glass-sponge spicules	Grass phytoliths
Iron					
Magnetite (Fe_3O_4)	Magnetosomes			Arthropods Molluscs Vertebrates	
Greigite (Fe_3S_4)	Magnetosomes				
Siderite ($FeCO_3$)	Extracellular precipitates				
Vivianite ($Fe_3[PO_4]_2 \cdot 8H_2O$)					
Goethite ($\alpha FeO \cdot OH$)			Extracellular precipitates	Chitons	
Lepidocrocite ($\alpha FeO \cdot OH$)			Extracellular precipitates	Chitons	
Ferrihydrite ($5Fe_2O_3 \cdot 9H_2O$)				Molluscs	Angiosperms
Manganese					
Manganese dioxide (MnO_2)		Intracellular or extracellular precipitates around spores			
Barium					
Barium sulfate ($BaSO_4$)		Algal-plastid gravity sensors Xenophyophore skeletons		Sense organs: statoliths (otoliths)	
Strontium					
Strontium sulfate ($SrSO_4$)		Acantharian tests		Mollusc shells	

Table 2: Biomineralization: Biologically Controlled Mineral Precipitation by Cells
Each of the five kingdoms contains organisms that can mobilize minerals and incorporate them into their bodies and support structures. A few examples are given here (see Lowenstam and Wiener, 1989, in Bibliography, page 209). (The aid of John Stolz and Heinz Lowenstam in developing this table is gratefully acknowledged.)

to eukaryotes: Protoctists, fungi, animals, and plants all merge their bodies, their sex cells, or at least their nuclei in the mating act. The survivor of the sexual fusion is the recombinant organism, which, if it remains healthy, later also reproduces.

Reproduction starts with at least one live organism, whereas sexuality nearly always begins with at least two, who find each other and mate. The result of sexuality may be only one recombinant offspring (as in bacterial conjugation or the *Chlamydomonas* zygote), two recombinants (as in ciliate mating, where both "parents" but no "child" emerge as recombinants after mating), or more (as with people, where there are at least three: the child—the recombinant—and the two parents). For the ciliate *Stentor coeruleus*, sex is lethal: The mating partners die three or four days after their thirty-six–hour embrace. Thus, *Stentor* always buds new organisms; it reproduces without sex. There are many examples of sex without reproduction; the association of sex with reproduction certainly is not a biological rule. Only in certain animals and plants does mating regularly lead to reproduction. Three protist individuals mating to produce only one healthy individual (as occurs sometimes) hardly qualifies as reproduction because the result is a decrease in the number of organisms. All organisms that survive require some sort of eventual reproduction, but one parent with many offspring is often the rule, and many species lack sex altogether.

Gender

A universal feature of sexuality in protoctists, fungi, animals, and plants is gender and the ability of complementary cells or organisms to recognize each other's gender. Recognition of gender (i.e., complementary mating type) is followed by nuclear exchange or cell fusion. Although gender-determining mechanisms differ widely, mature cells differing in gender are attracted to each other, their smembranes fuse, and the cells transfer either nuclei alone or nuclei accompanied by cytoplasm. These sexual acts are followed eventually by karyogamy, nuclear fusion. Karyogamy recombines, in a single nucleus, genes from different parents. The reiterative, cyclical nature of cell fusion (usually fertilization) and its relief (usually meiosis) are major characteristics that distinguish all sexual protoctists from both their prokaryotic antecedents and their nonsexual relatives.

Multicellularity

In the development of many organisms (examples exist in each of the five kingdoms), cells fail to separate after division, and multicellular beings form. Many kinds of multicellular bacteria, protoctists, and fungi are known, even though unicellular beings are common in these kingdoms, too. In animals and plants the single-cell stage is limited to the development of eggs and sperm. All members of these two larger kingdoms are multicellular throughout nearly all of their life history.

Whereas the junctions or connections between cells of plants involve holes through the plant cell walls (plasmodesmata), many kinds of junctions connect animal cells: gap junctions, desmosomes, septate junctions, and others. Protoctist cell connections vary but include plasmodesmata and pit connections, like those of fungi. Protoctists and fungi may have walls between their cells, a specific form of cell connection called septa.

Multicellular, even differentiated, organisms are known in all five kingdoms; therefore, the dichotomy between unicellular and multicellular does not help to define the kingdoms. Many lineages of single-celled bacteria and nearly all lineages of protists reached multicellular status independently from each other (e.g., cyanobacteria, myxobacteria, actinobacteria, ameba–slime molds, diatoms, chrysophytes, and so forth). Although members of the kingdom Fungi may be unicellular (e.g., yeasts), animals and plants, because they grow from multicellular embryos, are always multicellular. Unicellular animals and unicellular plants do not exist! In this book all traditional "protozoans" (so-called unicellular animals) and algae (so-called unicellular plants) are included in the protoctists.

Populations, Communities, and Ecosystems

Populations of organisms are simply members of the same species found in the same place at the same time, such as a herd of bison on the plains or all the daisies in a field. Communities, on the other hand, are composed of sets of populations of organisms of different species in the same place at the same time. The habitat scenes of this book illustrate communities. Communities form ecosystems such as ocean, forest, jungle, or desert. Together, all live organisms on the face of the Earth, from the bottom of the abyss to the air at the top of the atmosphere, make up the biota—all living matter on Earth at a given time. Paleontology is the study of the past biota; most species that ever lived on Earth are already extinct. The biota and all of its natural habitats together are called the biosphere.

A very active system, the biosphere consists of all of its ecosystems. All of Earth's ecosystems taken together and interacting as a planetwide system of life in its environment have certain physiological tendencies. The biota greatly influences its surrounding temperature and moisture content. In fact, we hypothesize that the biota helps to control the composition of the atmospheric gases and the temperature of air and water. By producing ammonia as waste, the biota lowers acidity. Organism growth and behavior may even help control ocean salinity. The tendency of the biota, the sum of life on Earth, to modulate conditions in its immediate environment has been called Gaia. More specifically, Gaia is the largest composite ecosystem of all, the sum of the smaller ecosystems on Earth, composed of all the life forms with their interactions tending to modulate their immediate environments. This book introduces the components of Gaia (all of our planetmates) about most of which we tend to be completely unaware.

The five-kingdom groupings here reflect as much as possible the reorganization of the biological sciences: molecular biological detail, ultra-structural and genetic analysis, and the revolution in paleobiology. The unambiguous classification of the diversity of life requires recognition of the key importance of microorganisms, especially the poorly known protoctists, in the evolution and present-day distribution of life on Earth.

Biomes and Planet Earth

Biomes are huge, contiguous territories that are geographically identifiable (e.g., the world ocean, the northern tundra, or the equatorial tropical rain forests). Biomes are composed of ecosystems; for example, coastal zones, open ocean, and abyssal plains are all part of the ocean biome. The sum of all biomes is the biosphere, which is the twenty-kilometer-deep, hollow sphere or spherical ring at Earth's surface in which life resides.

Systematics: Classification and Naming

Every society has a classification system that often is based on the organism's direct relationship to people (Figure 4). Our modern system of classifying organisms began with the Swedish scientist Carl von Linné (1707-1778), known by the Latin version of his name, Carolus Linnaeus. He revolutionized the naming and identifying of organisms by giving each type two names, a generic name and a specific name. Binomial ("two-name") nomenclature is the proper term for Linnaeus's system, which is still in use today. Every known type of organism is identified by its two-part technical name, always written in italics and usually derived from Latin or Greek. The first part of the name refers to the larger group (genus) to which an organism belongs. The second name refers to its species.

The genus to which most soup beans belong is *Phaseolus,* although the larger faba beans belong to the genus *Vicia.* Chimpanzees and gorillas belong to the genera *Pan* and *Gorilla* respectively. Chimps, gorillas, and people all belong to the mammalian order Primates. *Homo sapiens* refers to human beings; *Homo* is the genus name, and *sapiens* the species name (in Latin *Homo* means "human," and *sapiens* means "wise"). Species are grouped into genera, genera into families, families into orders, orders into classes, and classes into phyla. Each phylum can be assigned to one of the five kingdoms.

Dogs, for example, are canids; they belong to the kingdom Animalia, the phylum Chordata, the class Mammalia, the order Carnivora, and the family Canidae. Their genus is *Canis. Canis familiaris, Canis lupus,* and *Canis latrans* are the scientific names of the domesticated dog, the wolf, and the coyote, respectively. No matter what country naturalists or scientists are in, they use the genus and species names. An important benefit of taxonomy is universal scientific terminology. Even scientists writing with Japanese or Russian letters use the Latin alphabet for these scientific names. If the technical genus and species name is always the same, everyone knows without a doubt which organism is meant, even if it has a different common name in each country. Lightning bugs and fireflies are flying insects that glow in the dark. Are they the same? To be certain, one needs to use the scientific name.

The inadequacy of the two-kingdom system has been known since the last century, when the German scientist Ernst Haeckel (1834-1919) created three kingdoms: Monera, Plantae, and Animalia. All modern scientists agree that bacteria, fungi, and other microscopic organisms do not fit into a two-kingdom scheme, although the details of the overall taxonomy of life are still being worked on today. Our summary, of course subject to revision, is in the Appendix, which begins on page 165.

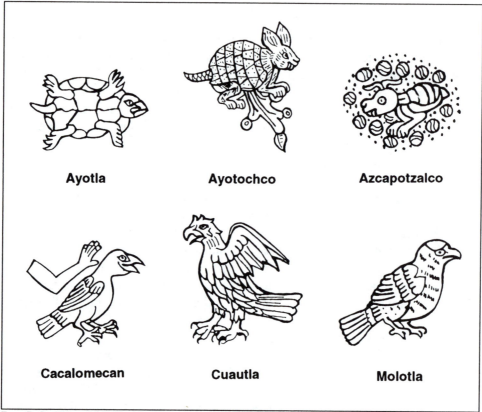

From R. L. Castro and F. Garrido. 1984. *Escribir con Imágenes: Antiguos Nombres Mexicanos para Iluminar.* Ediciones del Ermitaño, Tlacopec, Mexico.

Figure 4
Ancient Mexicans named all the organisms around them and specified their usefulness. The knowledge in the Nahuatl language was recorded in the sixteenth century by Fray Bernardino de Sahagún in the *Florentine Codex*: ". . . 2nd chapter, which telleth of all the different kinds of birds . . . 4th paragraph, which telleth of all the birds of prey. . . . The eagle is fearless, a brave one. . . . It can face the sun. . . . It is brave, daring, a screamer, a wing-beater. . . . it constantly grooms itself."

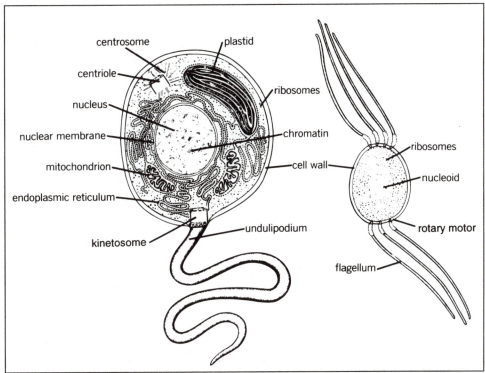

Christie Lyons

Figure 5
The bacterial cell (right) is nonnucleated, or prokaryotic. Its DNA is dispersed throughout the cell, and it has a flagellum for locomotion. By contrast, the eukaryotic cell (left) has a nucleus, which contains its DNA in the form of chromosomes. Its organelle of locomotion is the undulipodium. The details of the rotary motor flagellum and the far larger whiplike undulipodium are shown in Figure 2 (page 4).

Five Kinds of Life

Bacteria are strange. Electron microscopy has shown that not only are bacteria very different from plants, animals, and fungi, but they are also very different from each other and from the microscopic protoctists (the protists). All scientists put bacteria in their own kingdom: the Procaryotae, or Monera. Unlike all other organisms on Earth, bacteria lack nuclei in their cells (Figure 5). Also unlike the other kingdoms of organisms on the planet, all bacteria can, in principle, trade genes with each other. Still, even until the 1960s most biologists grouped bacteria together with all the other nonanimals; since bacteria were alive but were not animals, it was logical to call them plants. Haeckel's recognition of microbes as an altogether different kind of life is still having trouble catching on.

The first adoption of Haeckel's ideas in the United States came in 1956 when biologist H. P. Copeland (1902-1968) separated Haeckel's one kingdom Monera into two. He wrote a book presenting a four-kingdom classification of organisms—which nearly no one read at the time. Copeland placed bacteria into one kingdom and protoctists (which included the fungi) into another kingdom of microscopic beings. Copeland and other biologists realized that all organisms are made of cells that either have nuclei (eukaryotes) or do not have nuclei (prokaryotes). Improved microscopes made it easier to determine the presence or absence of nuclei. Since all plants and animals have nuclei, animal cells are much more like plant cells than they are like bacteria. Fungi are eukaryotes, too, because they all have nuclei in their cells.

By 1959 Cornell University Professor R. H. Whittaker (1924-1980) had thoroughly read Copeland's little book on the "classification of the lower organisms." Whittaker proposed a five-kingdom system that divided life forms into the groups that we present here. Whittaker, who studied pine forests in New Jersey and deserts in the southwestern United States, found bacteria and fungi to be so unlike plants that he could no longer call them plants at all. At the same time, his system does not split life into so many different kingdoms that one cannot remember them.

This book is based on Whittaker's five kingdoms: (1) Monera, (2) Protista, (3) Fungi, (4) Animalia, and (5) Plantae. These correspond to (1) bacteria; (2) algae, ciliates, and approximately forty other groups of microbes and related larger organisms that are not members of the plants, animals, or fungi; (3) molds, yeasts, and mushrooms; (4) animals; and (5) plants (bryophytes and tracheophytes). These are the five major kinds of life on Earth (Figure 6).

The only slight change we have made is to follow Copeland (in recognizing the great protoctist kingdom) rather than Whittaker. We recognize the limitation of multicellularity versus unicellularity and place all protists and their multicellular descendants into the more inclusive and now far better studied group: the kingdom Protoctista. This change allows us to include many larger nonanimal, nonfungal, and nonplant organisms in this most obscure kingdom. We can distinguish between the smaller members of the kingdom, the protists (single cells with nuclei), and the protoctists (many-celled organisms with nuclei but that do not conform to the descriptions of plants, animals, or fungi).

Multicellular animals used to be called metazoans, whereas "unicellular animals" were called protozoans. In the five-kingdom system, however, there is no such thing as a one-celled animal. In the five-kingdom system all animals grow from a mother's egg that has been fertilized by a swimming, smaller cell (called a sperm) from the father. The fertile egg develops into a blastula, which is a small, hollow ball of cells that usually goes on to develop a gastrula, an embryo with the beginnings of a digestive system that generally goes on to form tissues.

Size

As a group, members of the microbial kingdoms (bacteria, protoctists, and fungi) range enormously in size. From the smallest coccoids, which measure 0.09 micrometer in diameter, to the giant kelps, organisms that extend in size over eight orders of magnitude are included. Botanists and marine biologists do not like to call the gigantic members of the kingdom protists, a term with connotations of the very small. In the history of biology the term protist once included the prokaryotes. In this work the term protist is informal, referring to members of the kingdom Protoctista that require the use of a microscope to be visualized.

Bacteria on the Geological Time Scale

All organisms have bacterial ancestors. Prokaryotes first appear in the fossil record approximately 3,500 million years ago. From some protoctists, the first new organisms to form from these prokaryotes, emerged members of the other eukaryotic kingdoms: Animalia and Fungi from unknown heterotrophs, and

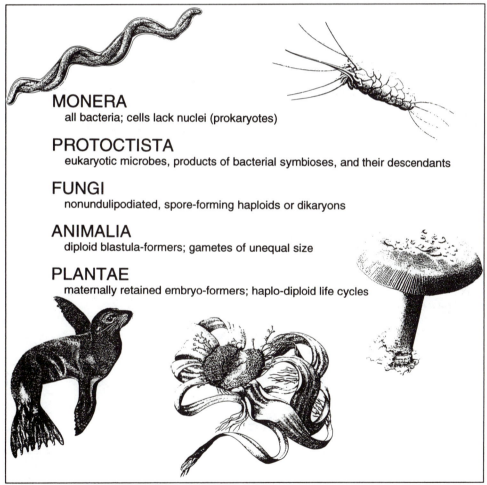

Christie Lyons

Figure 6

Five kinds of life. Clockwise from top left: *Cristispira* (spirochete), *Mallomonas* (golden-yellow alga), *Cantharellus cibarius* (chanterelle), *Welwitschia mirabilis* (gnetophyte), *Zalophus* (California sea lion). Protists are the smallest protoctists, from which all the other proctoctists, as well as fungi, animals, and plants, evolved.

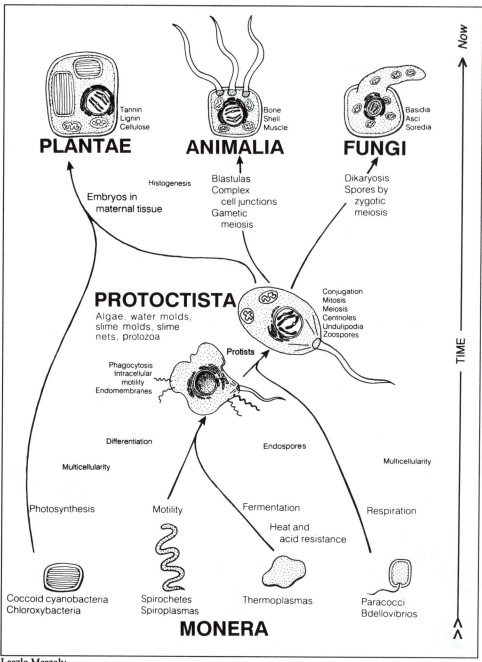

Laszlo Meszoly

Figure 7

This phylogeny, or family tree, of life on Earth indicates the ancient bacterial (prokaryotic) lineages, as well as those of their eukaryotic descendants that evolved by fusion of lineages (i.e., by symbiogenesis to form protists). The protist cells are ancestral to all other protoctists, as well as to the plants, animals, and fungi. This diagram thus depicts what is referred to as the serial endosymbiosis theory (SET) of evolution. The joining of arrows indicates the origin of cells by symbiogenesis.

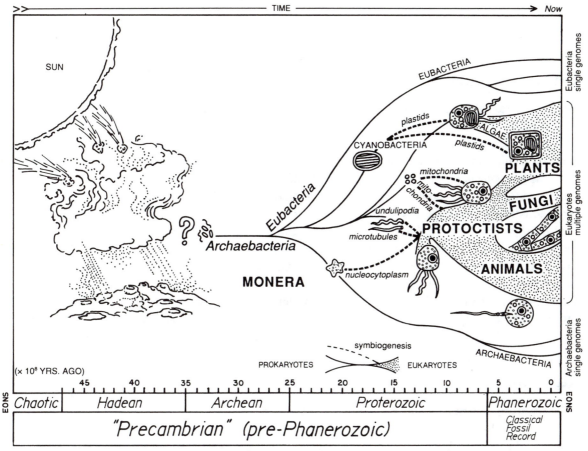

Kathryn Delisle

Figure 8
In this geological time scale phylogeny, the evolution of life is plotted across the geological time scale at the bottom. Eukaryotic kingdoms, those that evolved by symbiogenesis, are shaded. The symbiotic origins of eukaryotic organelles from bacteria are indicated by dashed lines.

Plantae from green algae (Figure 7). These three kingdoms, which in general appear later in the geological record than protoctists do, are defined in this book, and the dates of their earliest fossils are illustrated (Figure 8).

Animals (diploid organisms developing from blastular embryos) from the Ediacaran biota (see Figure 11, page 50) are found in the fossil record over 700 million years ago. Fungi (haploid and dikaryotic organisms developing from desiccation-tolerant spores and lacking undulipodia at all stages of their life cycle) appeared, primarily in association with plant roots, in the late Silurian or lower Devonian, some 450 million years ago. Plants (organisms that develop from embryos that are not blastulas and that are surrounded by maternal tissue, and in which the haploid alternates with the diploid generation) also appeared in the lower Paleozoic era. Fossils interpreted to be robust walled cysts of protoctists are recorded in the fossil record well over a billion years ago. *Grypania,* a coiled-tube fossil visible to the unaided eye, may be a two-billion-year-old photosynthetic protoctist.

Common Ancestry

Nutrition, sexuality, sensitivity, and social behavior vary enormously among members of the biota, yet all organisms alive today have many common characteristics. All are composed of the same types of protein, nucleic-acid, and lipid molecules and share the autopoietic organization discussed earlier (see page 1). The hypothesis of Charles Darwin (1809-1882), the English evolutionist, that all life on Earth has common ancestry is well documented by evidence from biochemistry and from the fossil record. Similarly, the claim of the great Russian geologist/mineralogist Vladimir I. Vernadsky (1863-1945) that all life on Earth today depends on light from the nearest star, the sun, and that all living forms are in physical contact with each other via the air and water is also strongly substantiated by modern science. Life is a unified phenomenon, a covering of the Earth that has persisted through nearly four billion years of Darwinian time and throughout the biospheric ring (ten billion cubic kilometers) of Vernadskian space.

Associations

All organisms depend on others to provide their food, remove their waste, and produce their breathable gas. No species can survive in the absence of others, although the details vary from organism to organism. Some plants require certain insect species for pollination, but the contact between the sexual parts of the plant and the pollinating hymenopteran (bee or wasp), for example, may be brief. Some protoctists (e.g., myxozoans and apicomplexans) derive all of their nutrition from members of different kingdoms and phyla; they are symbiotrophs. If organisms kill or seriously threaten their living habitat, they are called necrotrophs.

Symbionts are organisms of different species living in long-term association. The nature of symbiotic relationships changes from casual to permanent, from temporary to obligate, depending on many factors—of which environmental conditions and evolutionary history may be the most important. Symbiogenesis is the origin of new form and function that emerges from long-term symbiosis (e.g., the appearance of the rumen in the ancestors of cattle or of the trophosome in pogonophoran worms).

Chapter 1: Monera

The kingdom Monera comprises all the bacteria—the subvisible prokaryotic, or nonnucleated, organisms. Most are simple unicellular organisms, such as the omnibacteria. The most complex bacteria undergo developmental changes in form: Unicellular bacteria may come together in packets of millions and metamorphose into stalked structures that release resistant sporelike microcysts (e.g., Myxobacteria, see page 38). Some grow long, even, branched filaments; others form fat, sessile bodies that bud off swimming, flagellated, single-celled offspring (Figure 9). The distinctive type of movement displayed by flagellated bacteria is shown in Figure 10.

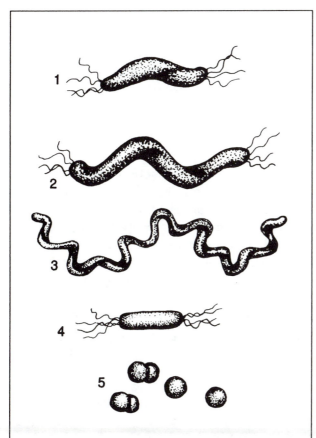

J. Steven Alexander

Figure 9
Bacterial shapes. (1) Vibrio (curved). (2) Spirillum (helical, external flagella). (3) Spirochete (helical, internal flagella). (4) Bacillus (rod-shaped). (5) Coccoid (spherical).

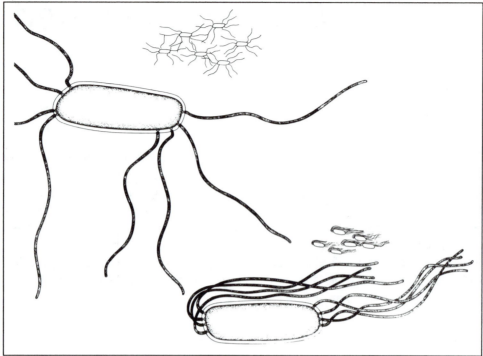

Christie Lyons

Figure 10
Bacterial flagella exhibit two basic types of movement: They twiddle (top left), in which the flagella move in an uncoordinated way and the bacterium remains stationary, and they run (lower right), in which the flagella move in a coordinated and synchronized way, propelling the cell in a certain direction (here to the left).

Noted for their complex chemistry, bacteria effect many chemical transformations. Metabolically, bacteria are far more diverse than all the eukaryotes (protoctists, fungi, animals, and plants) taken together. Some monerans—such as the methanogenic, halophilic, and thermoacidophilic bacteria—are genetically distinct enough to be considered a subkingdom: the archaebacteria. This distinction is based mainly on comparing the sequence of nucleotides in ribosomal RNA (rRNA). This universally distributed molecule is believed to have mutated relatively slowly through 3.6 billion years of evolutionary time; thus, its nucleotide sequence—based on comparison of bacterial rRNA to that of plants, animals, protoctists, and fungi—is relatively unchanged. By comparing the RNA sequences of different groups of organisms, we can estimate how closely related they are.

Although the nomenclature of bacteria is in dispute, more than 10,000 "species," including the cyanobacteria, have been named and described in the bacteriological literature. Many more are still unidentified; most bacteriologists agree that the majority of bacteria have not been grown and studied. Very few descriptions of bacteria in their natural communities are available.

Concerned with organisms important to medicine and agriculture, bacteriologists rely mainly on a practical manual, *Bergey's Manual of Systematic Bacteriology*. They do not conform their nomenclatural and taxonomic practices to those of other biologists. Thus, some of the groupings in this text are simplified from those found in the standard reference works (see Bibliography, page 209) so that the taxonomic level of phylum is conceptually comparable throughout the five kingdoms. We recognize seventeen prokaryotic phyla—fewer than those of the animals or protoctists but more than those of the plants or fungi. These phyla group the bacteria by easily distinguishable, ecologically important characteristics, both morphological and metabolic. Where molecular evolutionary information is available, we keep genetically related groups together. Unlike change through time in other organisms, in bacteria evolutionary change seems to be reversible. (In eukaryotes the appearance of new species seems to be an irreversible process.) Bacteria, which transfer their genes back and forth to other bacteria throughout the biosphere, group themselves into communities or guilds that form a global network, as Sonea and Panisset (1983) describe.

Pasture

Phylum: Methanocreatrices. To date, three great groups, or phyla, of archaebacteria have been recognized: salt-loving halophiles, hot sulfur/acid–tolerant thermoacidophiles, and these bacteria, Methanocreatrices, that make methane gas and live in the absence of oxygen. Some scientists believe that archaebacteria are the oldest form of life on Earth, which is why they were given a name that means "ancient bacteria." The archaebacteria in this phylum are called methanogens because they produce methane, the same gas that we pipe into our stoves for fuel.

Because they are poisoned by oxygen, methanogens are restricted to anoxic environments: sewage, marine and freshwater sediments, and the intestinal tracts of some animals, such as the rumen of the four-chambered stomach of cattle (*Bos taurus,* **A**). (The shaded area indicates where bacteria that can break down cellulose live.) Because they can act on only a limited number of substrates, methanogens depend on other bacterial fermenters to convert a wide range of organic compounds into the hydrogen and carbon dioxide gas that they need to produce methane. Of the 400×10^{12} to 600×10^{12} grams of methane released into the air each year, an estimated 74% comes directly from methanogens.

Many different methanogens reside as symbiotic bacteria that have become organelles in ciliates (see page 94). Such methanogenic ciliates live in anoxic sea water, muds, lake sediment, or sewage water. Each cell of *Plagiopyla frontata* and *Cyclidium porcatum,* for example, contains hundreds of methanogenic bacteria.

The phylum Methanocreatrices contains nineteen genera; fifty species have been named. Pictured here are *Methanobacterium ruminantium* (**B**, in cross section, dividing), *Methanogenium marirgen* (**C**), *Methanobacterium* (**D**), *Methanosarcina* (**E**), and *Methanogenium cariaci* (**F**).

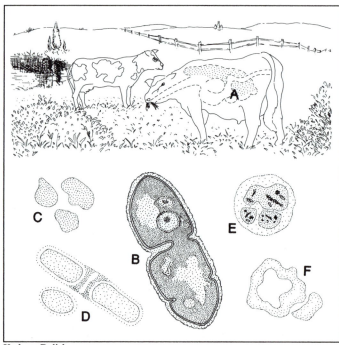

Kathryn Delisle

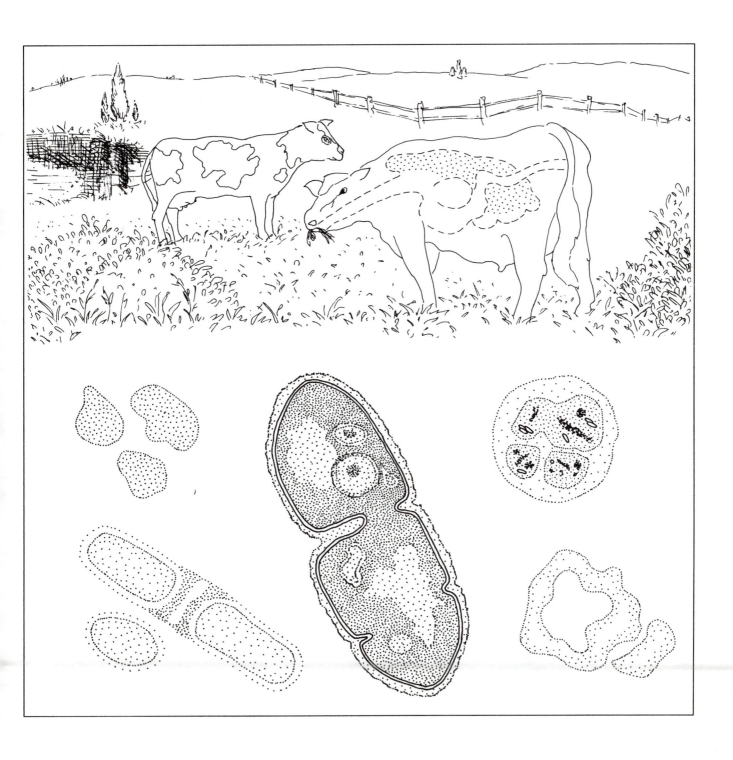

Hot Springs and Mud Flats

Phylum: Halophilic and Thermoacidophilic Bacteria. This phylum contains the two groups of archaebacteria that are incapable of methane production. Both groups thrive in extreme environments: The halophiles require salt to live, whereas the thermoacidophiles require hot, acidic conditions. Neither group forms spores.

Thermoacidophilic bacteria, such as *Thermoplasma acidophilum* (**A**), which lives in habitats like this hot spring in a geyser pool at Yellowstone National Park (**B**), grow vigorously at 60°C and at pH values from 1 to 2 (the pH of concentrated sulfuric acid). At room temperature they freeze and die immediately. Although *Thermoplasma* is a bacterium (and all bacteria are prokaryotes), its DNA, like that of eukaryotes, has a protein coat that probably protects its genes from the hot acid.

Halophilic bacteria, such as the spherical *Halococcus* sp. (**C**) and the rod-shaped *Halobacter* sp. (**D**), both much smaller than the salt crystals that surround them (**E**), live on mud flats in salt marshes (**F**). The cell membranes of halophiles are strong: Special components, such as derivatives of glycerol diether, allow them to withstand high salt concentrations. When exposed to low salt concentrations, however—even the fresh water of rain—many cells of halophiles burst open.

Because, like carrots, halophiles make pigments called carotenoids that color their cells bright pink or orange, at high population densities halophiles can be spotted from airplanes and orbiting satellites as pink scum on salt flats.

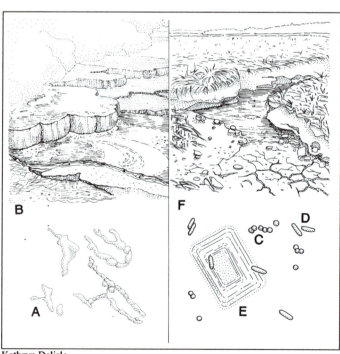

Kathryn Delisle

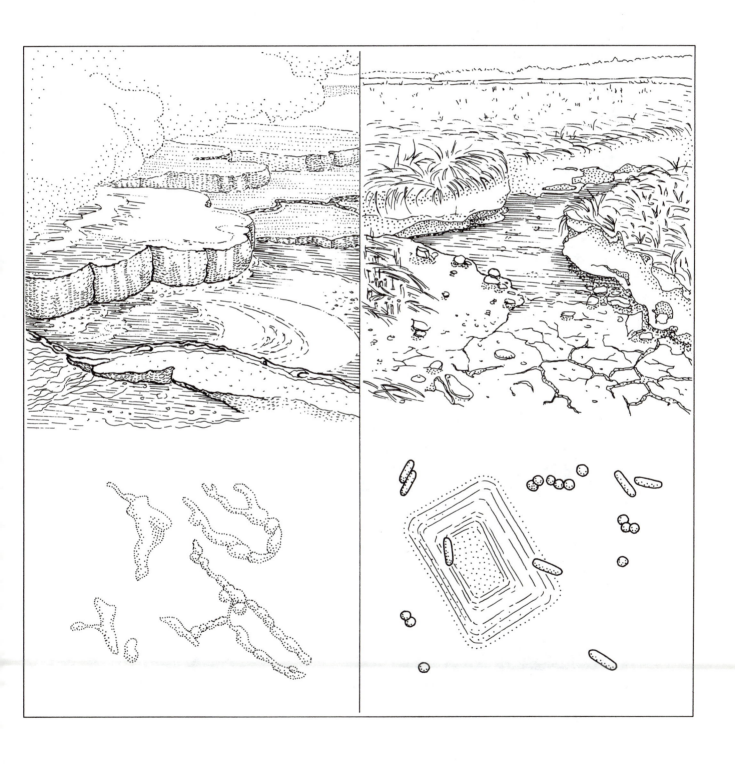

Barnyard

Phylum: Aphragmabacteria. Because of their chemical constituents, aphragmabacteria are classified as eubacteria rather than as archaebacteria. Most bacteria—in fact all except the three groups already depicted: the methanogens (see page 22), the halophiles, and the thermoacidophiles (see page 24)—are eubacteria. Because they lack cell walls, however, aphragmabacteria differ from all other eubacteria.

Like all other living beings, aphragmabacteria have flexible membranes made up of fat and protein structures, but they lack rigid walls outside of those membranes. They are therefore resistant to penicillin and other antibiotics that inhibit cell wall formation. When aphragmabacteria grow, their cells usually pile up on each other to form colonies of cells that look like fried eggs (**A**). Members of one group of aphragmabacteria, the mycoplasmas, have unusually high concentrations of cholesterol in their lipid membranes, suggesting a long

association between these bacteria and animal tissue rich in lipids. A cross section of a cell from the genus *Mycoplasma* (**B**) reveals its nucleoid, ribosomes, cell membrane, and fuzz on its surface.

Some aphragmabacteria can cause pneumonia in humans and birds and may even cause diseases in plants. *Spiroplasma*, another mycoplasma, can be isolated from diseased citrus tree leaves and grown alone in culture tubes in the laboratory, but whether it actually causes the disease is unknown.

Aphragmabacteria reproduce in several ways: They form small coccoids inside of parent cells (**C**), bud (**D**), or divide into two equal-size offspring by binary fission (**E**). Some aphragmabacteria have very little total DNA, about ten times less than other bacteria. These organisms--the tiniest, wall-less creatures--which lack any fixed shape, are the most minimal kind of life on Earth.

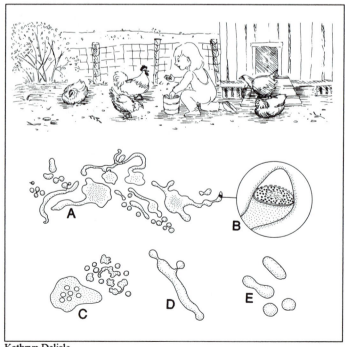

Kathryn Delisle

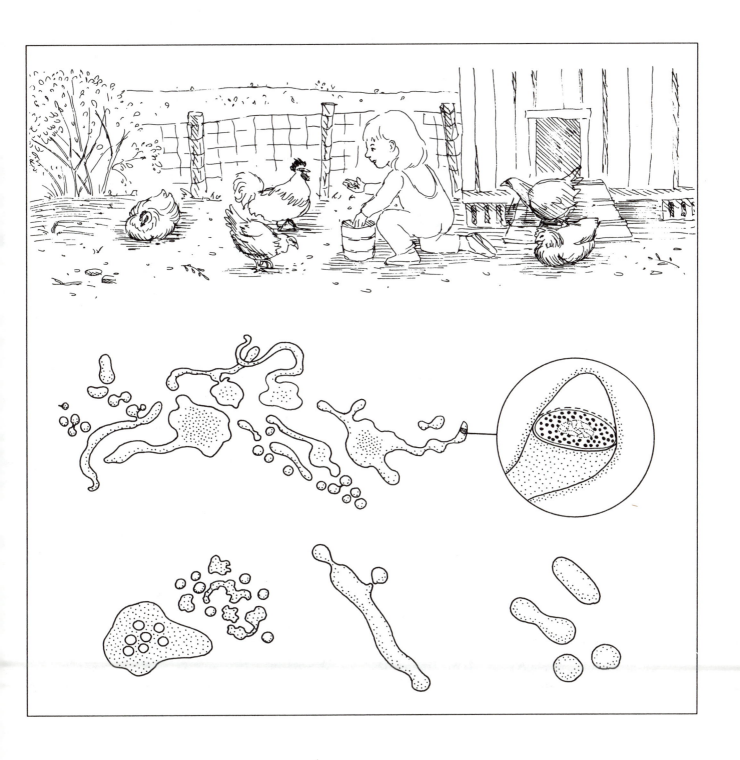

Clam Camp

Phylum: Spirochaetae. Spirochetes are bacteria that swim very rapidly, but unlike spirilla and other fast-moving bacteria, spirochetes have their flagella inside the outer membrane of their cell walls. Sometimes this design is referred to as the "snake-in-a-bag" arrangement. Some spirochetes are found in marine or freshwater environments; many are symbionts of animals. One type, *Treponema pallidum*, is found in great numbers in the genital sores of people who have symptoms of the venereal disease syphilis.

The three major groups of spirochetes—leptospires, spirochaetales, and cristispires—are pictured here. The leptospires (**A**, *Leptonema* or *Leptospira*) are one of the few types of spirochete that require gaseous oxygen. Certain leptospires are found in the kidney tubules (**B**) of mammals, such as raccoons and rats, which may or may not have a disease called leptospirosis.

Twelve different spirochaetales genera have been described, including the smallest, most familiar types like *Spirochaeta* (**C**). *Spirochaeta* is free-living and very similar in structure to the symbiotic *Treponema*.

Serpula is a similar spirochete that lives in the digestive system of pigs. The cross section of *Spirochaeta* (**D**) shows the flagella between the inner and outer cell wall membranes of these typical Gram-negative bacteria.

The cristispires (**E**) are much larger than the leptospires or the spirochaetales and are all symbiotic with marine molluscs (see page 142), such as oysters and clams. Cristispires inhabit a large gelatinous translucent structure called the crystalline style at the front of the stomach of the mollusc (**F**). The style, which helps molluscs grind algae for food, provides food and a haven for large numbers of the huge spirochete *Cristispira,* as well as for two smaller types of bacteria (a mycoplasma and a spirillum). *Cristispira* has as many as 300 flagella.

Other spirochete genera include *Borrelia* (in ticks and mammals, some associated with Lyme disease), the large spirochetes that live in the intestines of termites: *Clevelandina, Diplocalyx, Hollandina,* and *Pillotina*; the small *Mobilifilum*, with ten flagella; and the huge *Spirosymplokos*, with a unique composite structure.

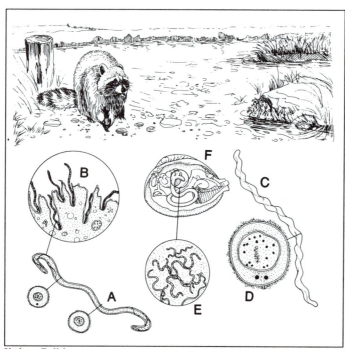

Kathryn Delisle

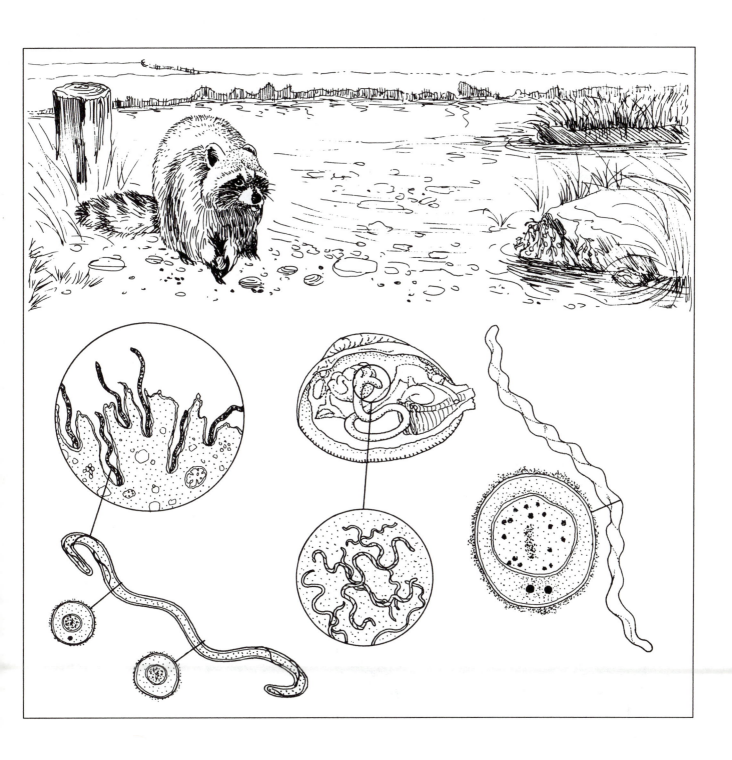

Rocky Brook

Phylum: Thiopneutes. Thiopneutes feed on organic matter, but instead of breathing oxygen and producing water as humans do, they take in elemental sulfur or sulfate and produce hydrogen sulfide. Just as with oxygen, however, taking in sulfate involves the disposal of waste hydrogen atoms. Thus, members of this phylum, such as *Desulfovibrio* (**A**), are important in the worldwide movement of the sulfur compounds that are necessary for their own proteins and those of all other organisms.

Shaped like commas, these Gram-negative bacteria invariably are found in the muds of sulfur-rich environments. The longitudinal section of *Desulfovibrio* (**B**) shows its flagella, ribosomes, cell wall, and nucleoid, as well as a comma-shaped structure associated with flagellar motility called the polar membrane. Many thiopneutes release pungent gases like hydrogen sulfide (H_2S), which smells like rotten eggs, making it easy to detect their presence indirectly. In iron-rich water, the hydrogen sulfide formed by these bacteria reacts with the iron, leading to the deposition of pyrite (iron sulfide, also known as fool's gold, **C**).

Phylum: Aeroendospora. To belong to this phylum, a bacterium must be at least a part-time aerobe, and it must produce endospores, which are resistant propagules that permit the distribution of these bacteria to all sorts of dry, barren places that the more sensitive growing cells cannot survive. The most notable genus of aerobic, endospore-forming bacteria is *Bacillus* (**D**), to which at least forty distinguishable kinds of rod-shaped bacteria have been assigned, making the bacillus group more diverse than some entire animal phyla.

A *Bacillus* parent cell produces only one airborne spore, which may land anywhere; in this dormant stage, bacilli can survive for years without water and nutrients. The *Bacillus* cells at bottom right (**E**) are depositing manganese. One bacillus species, *B. anthracis,* is associated with symptoms of a serious human lung disease called anthrax. This airborne disease caused about 60,000 deaths per year during the Middle Ages, especially among people who worked with domesticated animals. (A certain life stage of *B. anthracis* may grow in the intestines of sheep and goats, just as its relative, the fermenting bacterium *Arthromitus,* grows attached to the intestinal wall of wood-eating termites.) Today anthrax still affects 20,000 to 100,000 people a year, most of whom work with farm animals.

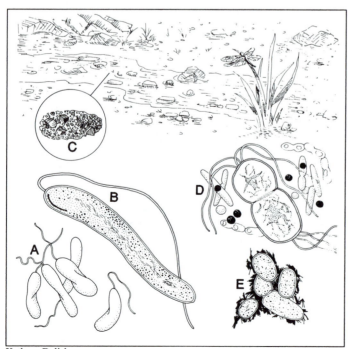

Kathryn Delisle

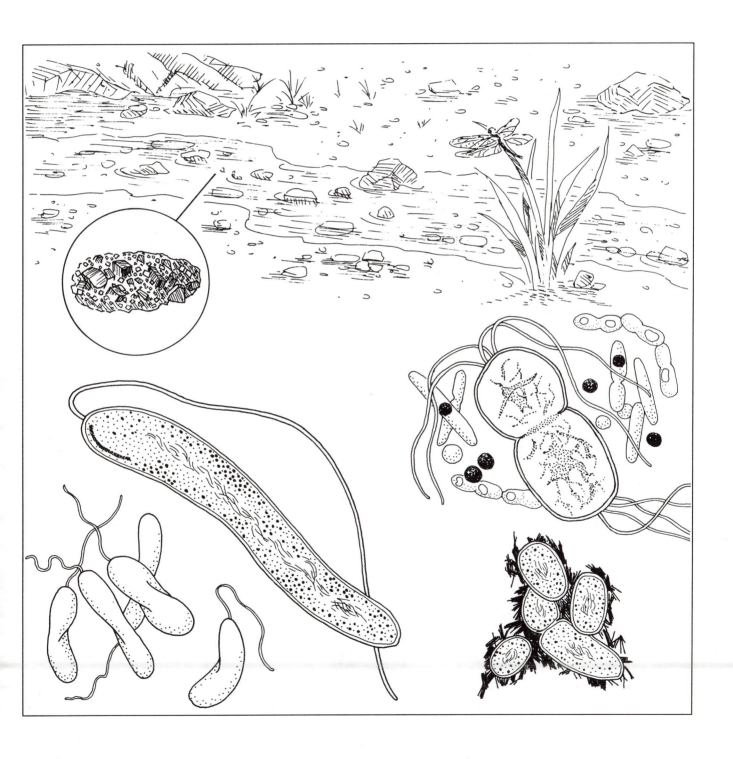

Pond Scum

Phylum: Anaerobic Phototrophic Bacteria. The bacteria in this phylum are phototrophs, primary producers in the world ecosystem. They derive their energy directly from the light of the sun and their carbon from the carbon dioxide of the air. Most require no organic compounds for food. There are three main groups of anaerobic photosynthesizers: green sulfur bacteria, purple sulfur bacteria, and purple nonsulfur bacteria.

Both green and purple sulfur bacteria (generally found living above the green sulfur bacteria in layers of mud) require sulfur, usually in the form of hydrogen sulfide (H_2S), as a source of hydrogen atoms so that they can reduce carbon dioxide into foodstuff carbon. The purple nonsulfur bacteria, on the other hand, use small organic molecules such as pyruvate or lactate as hydrogen donors.

The phototrophic bacteria are morphologically diverse. Some are single cells that swim; others are immotile. Some group together in packets, and others, like the purple sulfur bacterium *Rhodomicrobium vannielii* (**A**), named after the great Dutch-American microbiologist Cornelis van Niel (1897–1985), form stalked budding structures. Here the septa that divide individual cells can be seen (**B**). The cross section of *R. vannielii* (**C**) shows what it looks like under an electron microscope. Its cell wall, cell membrane, nucleoid with DNA fibrils, ribosomes, and thylakoid membranes are visible.

Some phototrophic bacteria form extensive sheets of cells in which the spaces between cells are filled with coverings, or sheaths, made of a gelatinous mucous material. The pond scum shown here (**D**), is a sign that prodigious numbers of phototrophic bacteria are present.

Phylum: Pseudomonads. The pseudomonads are a group of ubiquitous Gram-negative, flagellated, rod-shaped bacteria that have an astounding ability to break down organic compounds of all kinds, including organic ring compounds such as those in petroleum. The cross section of *Pseudomonas multivorans* (**E**), one species from the genus for which this phylum was named, shows its cell wall, cell membrane, nucleoid, DNA fibrils, ribosomes, and flagellum.

Members of the genus *Xanthomonas* (**F**), are serious plant pathogens, causing the rapid wilting and death of many food plants under conditions optimal for the bacterium. Other genera of the phylum include *Zoogloea* and *Bdellovibrio*. *Bdellovibrio* is a small, predatory bacterium that penetrates and enters the periplasm of many other Gram-negative bacteria. An aerobe, it reproduces in the periplasm of its victim by using the victim's cell material to make more of itself. After the victim's resources are exhausted, many new, healthy bdellovibrios burst out of its diminished remains and swim away to find new victims.

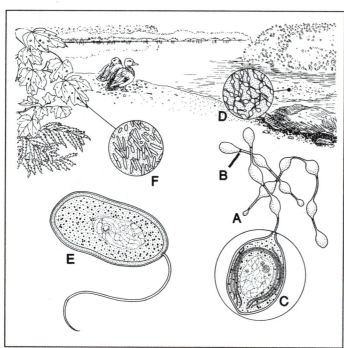

Kathryn Delisle

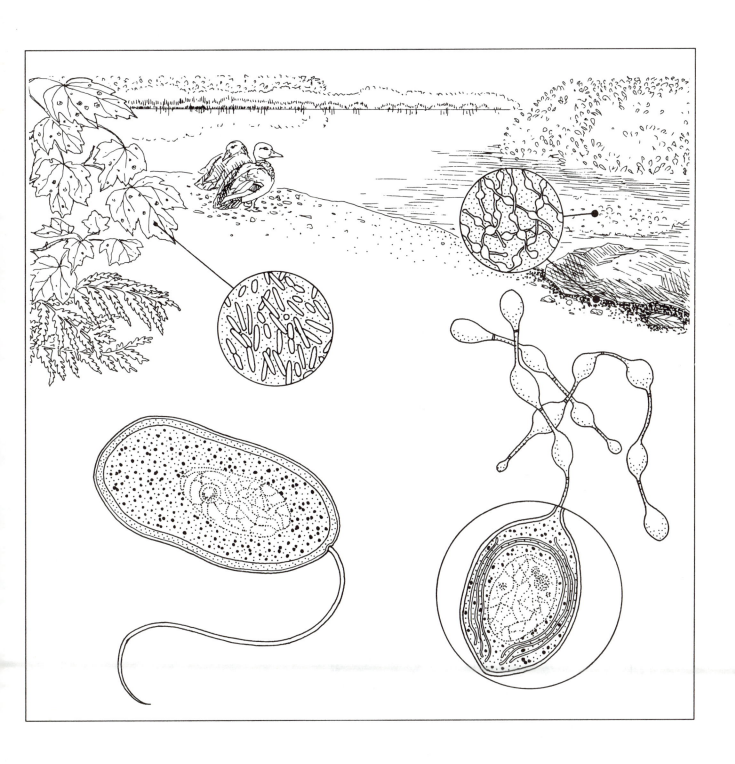

Salt Marsh

Phylum: Cyanobacteria. Together with the anaerobic phototrophic bacteria (see page 32), the protoctistan photosynthesizers (algae), and the photosynthesizers that bear embryos (plants), cyanobacteria sustain all life on Earth by converting solar energy and carbon dioxide into food. Unlike the bacteria that produce sulfur and sulfate as waste products of photosynthesis, however, cyanobacteria release the gas oxygen (O_2) into the air (just as plants and algae do).

During the Proterozoic eon (2,500 million to 600 million years ago) a diversity of bacteria, the most productive of which were the phototrophs, produced layered sediments. These muddy and sandy striped sediments, preserved as layered rocks that are interpreted to be fossils produced by the metabolic activity of microorganisms, are called stromatolites (**A**). Live cyanobacterial communities that grow in layered sedimentary structures like these are called microbial mats (**B**). Today microbial mats are found in the Persian Gulf, the Bahamas, Western Australia, northeastern Spain (Catalonia), the west coast of Mexico, and other seaside areas in semitropical and tropical regions.

Some of the most important cyanobacteria that make up these communities are *Spirulina* (**C**), *Oscillatoria limnetica* (**D**, with sulfur globules on its surface), and *Microcoleus* (**E**), in which many filaments (called trichomes) are inside a common covering (called a sheath). The large-filament *Johannesbaptista* (**F**) has coin-shaped cells, and the cells of the coccoid *Gloeothece* (**G**) are inside a spherical sheath. In many of these productive microbes the sheath acts like a pair of sunglasses, protecting the cells from fierce sunlight, yet letting enough light in so that photosynthesis can go on.

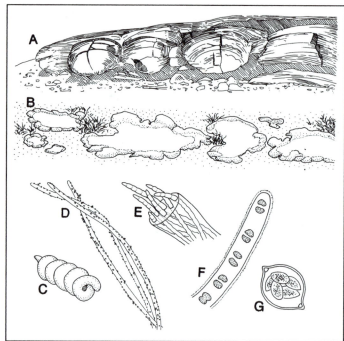

Kathryn Delisle

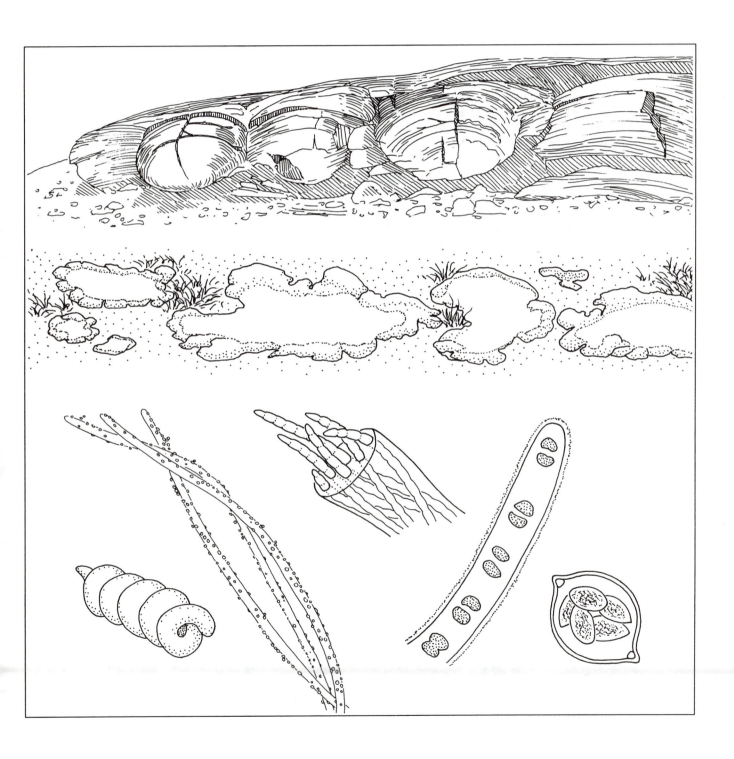

South Pacific Coral Reef

Although once thought to comprise a phylum of their own, the chloroxybacteria have been shown to belong to a subphylum of the cyanobacteria (see page 34), which they resemble in every way except in their pigments. Very few genera of the chloroxybacteria (also called Prochlorophyta) are known, since these bacteria were discovered only in the 1960s. The three genera that have been described are *Prochlorococcus, Prochloron,* and *Prochlorothrix.*

Chloroxybacteria are photosynthetic bacteria that, in structure and habit, greatly resemble the cyanobacteria. Unlike other bacteria but just like plants, however, these organisms have bright green chlorophyll: chlorophyll *a* and chlorophyll *b*. They lack both the blue-green pigments and the phycobiliproteins that make up the structures called phycobilisomes, which are found in all other cyanobacteria. Although the details of their photosynthetic apparatus are very similar to those of the chloroplasts of green algae and plants, the rest of their features--in particular the presence of a bacterium-style cell wall--place these prokaryotes squarely with other Gram-negative bacteria within the vast group of cyanobacteria.

Prochloron (**A**) is a simple nonmotile coccoid that lives as a symbiont in the walls of the cloaca (the combined excretory and reproductive canal at the end of the intestine) of sessile chordates called tunicates (**B**; see page 142). Although first discovered in the South Pacific, *Prochloron* also grows in tunicates near the marine station of the University of Arizona (at Puerta Peñasco) on the Gulf of California. The tunicate shown here, *Lissoclinum,* is in its natural habitat, a coral reef. In cross section, the cloacal cavity is visible (**C**). The cloacal wall (**D**) is embedded with bright green *Prochloron* symbionts.

Apparently *Prochloron* is an obligate symbiont because no one yet has been able to grow it separately. *Prochlorothrix,* however, is free-living; it grows easily in pure culture. Shaped like a filament, *Prochlorothrix* resembles the cyanobacterium *Oscillatoria* to the uninitiated. The only known species, which was discovered as a pollutant in a Dutch lake near Amsterdam, is called *P. hollandica.*

Still another green chloroxybacterium is found in great abundance in the world ocean. At the bottom of the photic zone, as far down as 200 meters of clear water, tiny coccoid bacteria thrive. Many ocean-going research ships have discovered these productive phototrophic bacteria and called them "unidentified green coccoids."

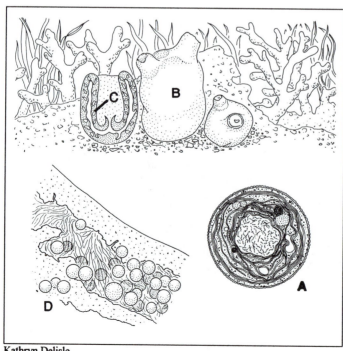

Kathryn Delisle

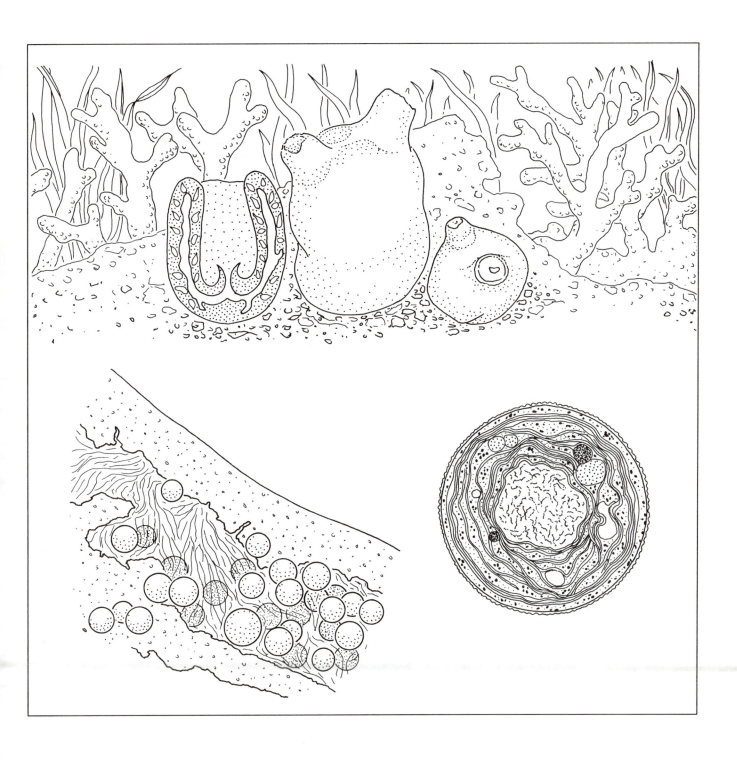

Garden Soil

Phylum: Nitrogen-fixing Aerobic Bacteria. The best-known nitrogen fixer is *Rhizobium,* which forms a nitrogen-fixing symbiosis with the root hairs of plants from the pea family (Leguminosae), such as soybeans, alfalfa, clover, and lentils. These bacteria provide one of the crucial elements for plant growth, nitrogen, by fixing the inert gaseous nitrogen (N_2) of the air and transferring it to the carbon compounds of the bodies of organisms (i.e., converting it into organic nitrogen, R-NH$_2$; "R" represents the organic part of the molecule).

Nitrogen fixation requires the complicated enzyme complex nitrogenase. Composed of at least two different kinds of proteins that contain iron and molybdenum ions, nitrogenase is highly sensitive to oxygen. Without the oxygen-sensitive nitrogen-fixing capability of these bacteria, which must have an ancient and venerable history, we would all starve from protein deficiency because of the lack of available protein-forming nitrogen.

Other genera of nitrogen-fixing aerobic bacteria include *Azomonas* and *Azotobacter* (**A**). The cross section (**B**) shows tubules within *Azotobacter,* which are unusual in bacteria and may be involved in cell division. Respiratory membranes, ribosomes, cell membrane, cell wall, nucleoid, and division furrow can also be seen. Nitrogen fixers are motile, but *Rhizobium* transforms: The bacteria lose their flagella, and their enzyme systems swell up when they are inside the roots of legumes. They become permanently embedded in the plant cells, forming nodules on the root as the symbiosis matures.

The nodules, often pink because of the presence of the oxygen carrier molecule leghemoglobin, stud the roots. They are seen easily with the unaided eye if the roots are removed carefully from the soil.

Phylum: Myxobacteria. In Greek *myxo* means "slimy." The myxobacteria, aerobic organisms that thrive in soil, are among the most morphologically complex of all bacteria. They glide by a poorly understood motion: Each cell moves slowly and is always in contact with a solid surface during movement. Individual cells come together to form colonies, and as water is depleted, they form differentiated, propagule-bearing stalks similar to those formed by the slime molds (see page 60).

Stigmatella aurantiaca (**C**) forms a distinctive tree-like structure with swellings (**D**), inside of which are bacterial myxospores. All myxobacteria are obligate aerobes with heterotrophic metabolism. They digest proteins and fatty-acid esters, and some probably can attack the lignin and cellulose of wood. They also digest the debris of other bacteria and of protoctists by secreting digestive enzymes into their surroundings and then absorbing the organic nutrients.

Phylum: Micrococci. These Gram-negative bacteria are spherical cells (*coccus* is the Greek for "berry") that characteristically divide to produce tetrads. They are either strictly or facultatively aerobic. Seven genera, distinguishable by morphology and arrangement of their cells, have been assigned to the phylum: *Micrococcus, Planococcus, Aerococcus, Sarcina, Gaffkya, Paracoccus,* and *Staphylococcus.*

Resistant to radiation, *Micrococcus* (**E**) grows in the coolant of nuclear reactors and resists ultraviolet and gamma rays. In a cross section of *Micrococcus* cells (**F**), the cell wall, membrane, ribosomes, and nucleoid are visible. Skin infections of humans and other animals are often associated with the growth of *Staphylococcus.* One strain is found in prodigious numbers in women suffering from toxic shock syndrome.

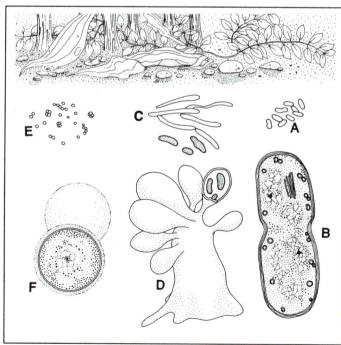

Kathryn Delisle

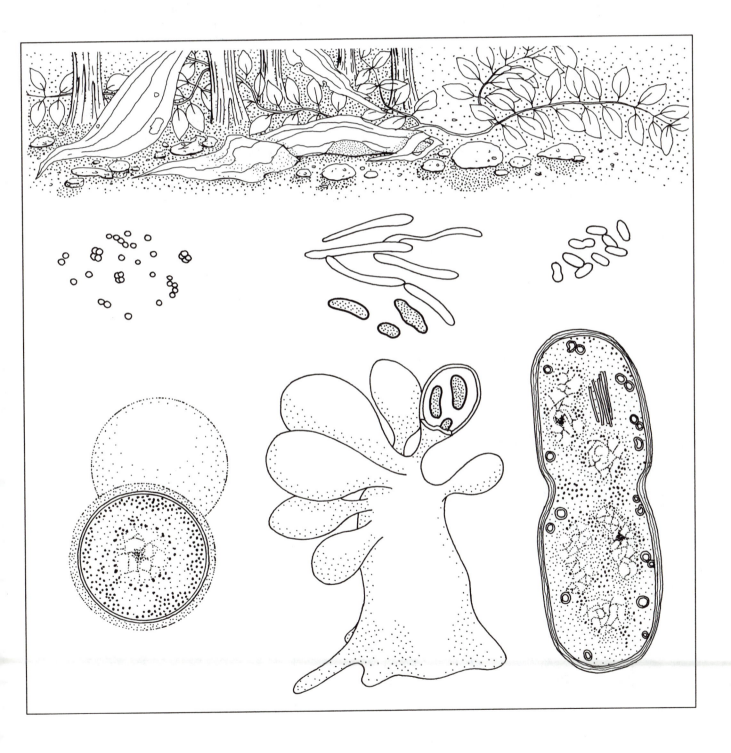

Lake Shore

Phylum: Omnibacteria. Most organisms in the world are omnibacteria. This vast group includes all the facultatively aerobic, Gram-negative bacteria that cannot use carbon dioxide or other inorganic compounds as food. In other words, omnibacteria are heterotrophs. In the absence of oxygen these bacteria do not stop growing; rather thsey continue to respire without oxygen.

They take in ions such as nitrate (NO_3^-), which, when it receives hydrogen atoms, is called the terminal electron acceptor, and they produce nitrogen gas (N_2), nitrous oxide (NO_2, also called laughing gas), or even ammonia (NH_3). Thus, omnibacteria are called facultative nitrate reducers, or denitrifiers. The latter expression comes from the disappointment of agriculturalists, who pay for added nitrate to fertilize their plants only to find that populations of omnibacteria have denitrified the useful nitrate (i.e., converted it to atmospheric gas) before the plants could take it up.

Examples of omnibacteria are *Caulobacter* (**A**, a swarmer cell and a cell with attached stalk); *Escherichia coli* (**B**), a resident of human large intestines and probably the most studied organism on Earth (shown here ready to divide); and *Aeromonas punctata* (**C**), pictured with *Hydra viridis* (**D**), a freshwater coelenterate (see page 122) with its many tentacles attached underwater to grass. This *Aeromonas* is a hydra symbiont that stays close to a good source of food: the algae-filled gastrodermal (digestive) cells of the green cnidarian in which it lives. Many types of aeromonads exist. They are abundant and common in ponds, lakes, and reservoirs.

The enterobacteria, a group of omnibacteria, are distinguished from one another by the carbohydrates that they can attack. Pathogens in this phylum include *Vibrio*, which induces diarrheal symptoms of cholera when it attaches to the intestinal linings of humans, and *Neisseria,* a bacterium found in great numbers in people suffering from gonorrhea or meningitis.

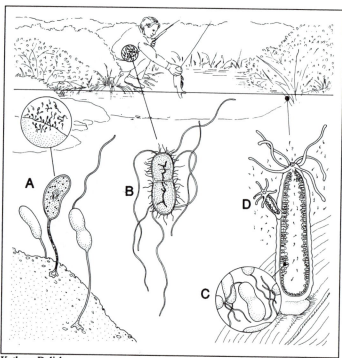

Kathryn Delisle

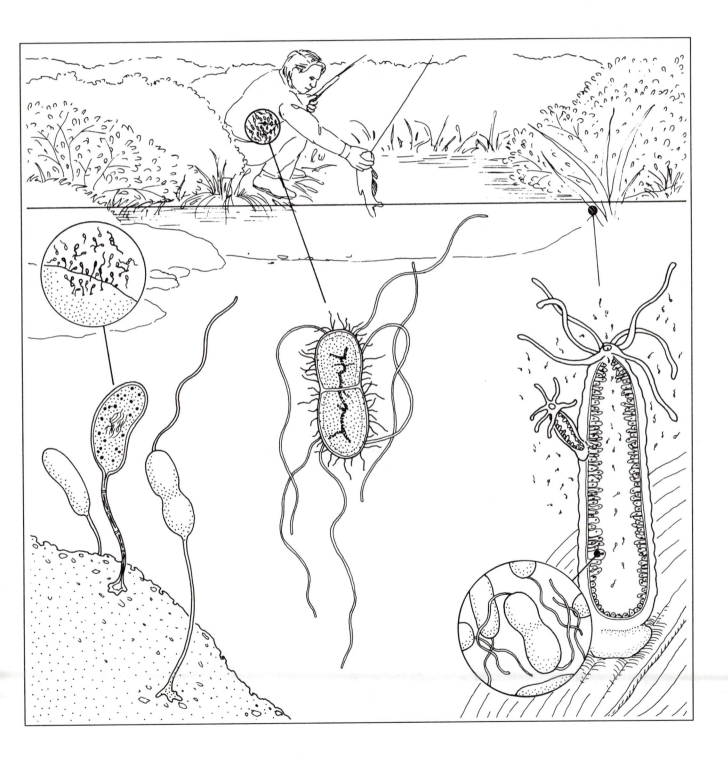

Ocean Edge

Phylum: Chemoautotrophic Bacteria. Chemoautotrophic bacteria are amazing because they live in water on inorganic chemicals alone. They need neither light nor food (organic compounds). There are several types of chemoautotrophs classified by the compounds that they oxidize to gain energy: nitrogen compounds, sulfur compounds, or methane.

Of the nitrogen oxidizers, *Nitrobacter*, *Nitrospina*, *Nitrocystis*, and *Nitrococcus* oxidize nitrite (NO_2^-) to nitrate (NO_3^-), whereas *Nitrosomonas*, *Nitrosospira* (**A**), *Nitrosococcus* (**B**), *Nitrosolobus* (**C**), and *Nitrosovibrio* (**D**) oxidize ammonia (NH_3) to nitrite (NO_2^-). *Thiobacillus* oxidizes reduced sulfide compounds containing sulfide (S^{2-}), sulfite (SO_3^{2-}), thiosulfate ($S_2O_3^{2-}$), or polythionate (a larger, more complex sulfur compound) to sulfate (SO_4^{2-}). Because the resulting sulfate mixed with water makes sulfuric acid (H_2SO_4), *Thiobacillus* of-

ten greatly acidifies its environment. Methylomonads use methane (CH_4) or methanol (CH_3OH) as their sole source of energy and carbon. They cannot even grow on food (i.e., complex organic compounds).

The chemoautotrophic bacteria are ecologically significant because they cycle elements critical to the growth and reproduction of living organisms: inorganic nitrogen and carbon compounds. *Nitrobacter winogradskyi* (**E**), a bacterium in the waters of the marine littoral zone, is shown here as it would look at very high magnification (with an electron microscope). The cell wall, respiratory membranes, nucleoid, ribosomes, and carboxysome (the storage granule of the CO_2-fixing enzyme) can be seen. Out in the water, where photosynthetically produced oxygen from above meets ammonia from below, these bacteria thrive.

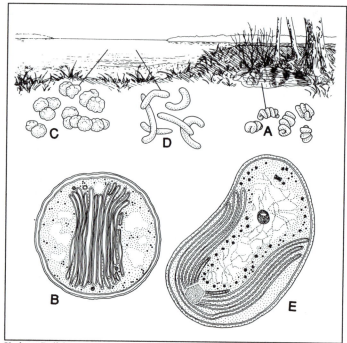

Kathryn Delisle

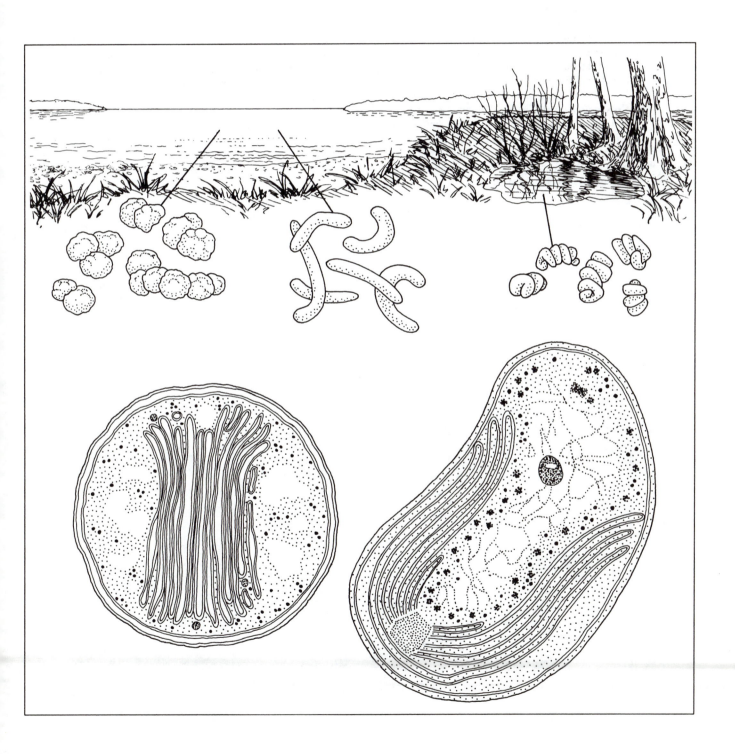

Foods

Phylum: Fermenting Bacteria. "Ferment" means to convert from sugar (such as glucose or sucrose) or fibrous substances (such as cellulose fibers or starch) to other organic matter—most notably ethyl alcohol, methanol, or even lactic or propionic acid. In short, fermentation is a process that occurs in the absence of oxygen in which carbohydrate is converted to alcohol or acid. Even though the enzymes involved in fermentation can be extracted from the cell, large-scale fermentation requires metabolizing microbes to assure the quality and quantity of the products.

This phylum contains fermenting geniuses such as *Clostridium*, which is used in processing linen fibers, and *Lactobacillus* and *Streptococcus*, both crucial in the production of yogurt and other milk products. Some of these specialized fermenters are able to break down almost anything (except plastic) through their anaerobic, fermentative metabolism.

Lactobacillus (**A**) from yogurt, spoiled milk, and the human mouth is grown easily in culture. In this cross section the sheath (capsule), outside cell wall, ribosomes, and DNA fibrils are visible. Different species of *Clostridium* cause gas gangrene and botulism. The lactic acid bacteria, such as *Lactobacillus, Streptococcus,* and *Leuconostoc,* are famous for their ability to ferment sugar (in particular that in milk) and to produce lactic acid, as well as formate, succinate, carbon dioxide, and ethanol. One product of this process is kefir, a carbonated yogurtlike drink that has been known in eastern Europe and western Asia for thousands of years.

Approximately twenty-five different bacteria and four yeasts are associated into a kefir curd (or granule) that acts as an individual, dividing and growing. The kefir curd, which looks like a granule of cottage cheese, is an example of "domesticated microbes."

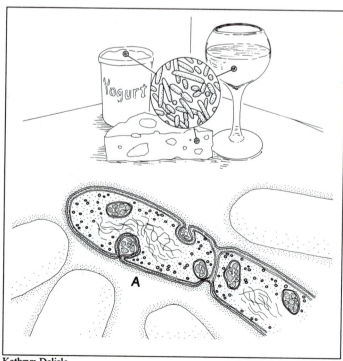

Kathryn Delisle

Woodland Stream

Phylum: Actinobacteria. Since many actinobacteria grow as stringy masses in colonies and appear fuzzy to the naked eye, superficially they resemble fungi. They live mainly in the aerobic zone of the soil and traditionally have been called actinomycetes (literally "ray-shaped fungi"). Electron microscopic and other studies show unequivocally, however, that they are bacteria and are not at all related to fungi. Although some actinobacteria, such as *Streptomyces,* have given us as many as 4,000 different antibiotics, others are considered disease agents because they are associated with symptoms of diphtheria, tuberculosis, and Hansen's disease (leprosy).

One notable genus of actinobacteria is *Frankia* (**A**), which forms nitrogen-fixing nodules on tree roots (**B**) similar to those found in *Rhizobium*-laden roots of legumes. Known as actinorrhizae, organisms like *Frankia* are common in the roots of certain temperate forest trees, including the alder (*Alnus,* **C**). Another actinobacterium, *Cellulomonas* (**D**), is capable of breaking down cellulose from trees. In this picture, offspring cells are still attached to two cells that have divided, showing the Y and V configuration typical of many actinobacteria, especially those in the family Nocardiaceae.

The odor of fresh earth, whether from newly hoed compost or from a rotten log recently overturned, comes from thriving populations of actinobacteria in their favorite habitat: organic-rich soil.

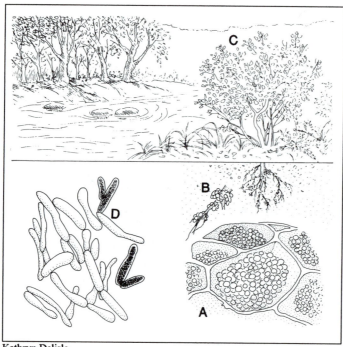

Kathryn Delisle

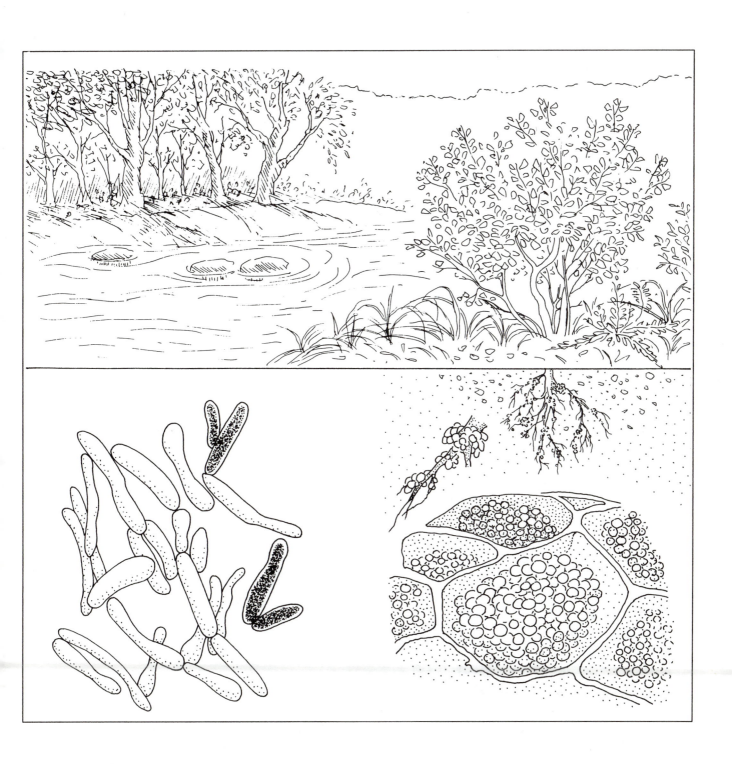

Chapter 2: Protoctista

The kingdom Protoctista includes organisms whose cells originally formed by bacterial symbioses. Protoctists are defined by exclusion: Members of this kingdom are not animals (which develop from an embryo called a blastula), nor plants (which develop from an embryo that is not a blastula and that is retained in the mother's tissue), nor fungi (which lack undulipodia at all stages and develop from spores), nor monerans (which have bacterial cell structure). All protoctists are nucleated and live in aquatic habitats. Eukaryotic microorganisms and their immediate descendants are protoctists: all algae (including the seaweeds), undulipodiated (motile) phytoplankton and water molds, the slime molds, slime nets, and that misnamed miscellany, the protozoans. Most protoctist cells have several eukaryotic characteristics, including nuclei. Most protoctists, but certainly not all, are also aerobes that respire oxygen in their mitochondria. Almost all of the swimming protoctists have [9(2)+2] undulipodia (see page 4) at some stage of their life history. In all algal protoctists the chlorophylls inside the cells are contained within membrane-bounded structures called plastids.

Some protoctists are marine, some live primarily in fresh water, and many inhabit the watery tissues of other organisms. Nearly every animal, fungus, and plant has a protoctist associated with it at some time during its development. Some protoctist phyla—such as the Apicomplexa (see page 98), the Microspora (see page 58), and the Myxozoa (see page 56)—include hundreds of species, all of which live necrotrophically on other organisms.

The protoctists show remarkable variation in cell organization, pattern of cell division, and life history. Like plants, some are photoautotrophs, which eliminate oxygen gas in the light; others ingest or absorb nutrients or living, swimming food heterotrophically. (Protoctists may be phagotrophs or osmotrophs or both at the same time). In many species, the type of nutrition depends on environmental conditions: When light is plentiful, they photosynthesize; in the dark, they feed. Although protoctists are far more diverse in ecological niche and nutrition than are animals, fungi, or plants, they are far less diverse metabolically than are the bacteria. That the appearance of protoctists may have occurred about two billion years ago is suggested by well-dated fossils such as *Grypania*, probably an alga. Protoctists subsequently diversified into a variety of weird beings, most of which are probably extinct (Figure 11). Fossil protoctists, especially thick-walled resting stages or cysts, can be extracted from shale rock treated with hydrofluoric acid.

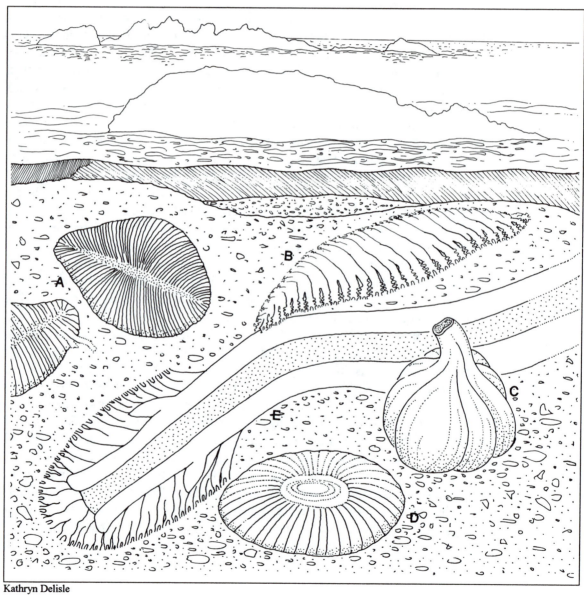

Kathryn Delisle

Figure 11
The fossil remains of the Ediacaran biota, first discovered in South Australia, date back to 600 million years ago. When first discovered, in the late 1950s and early 1960s, these organisms were called fauna and labeled as animals. Now some scientists believe that they were protoctists, some of which may have been ancestral to certain animal phyla. Pictured here in their shallow coastal habitat, at about half their natural size, are *Dickensonia* (**A**), an unnamed spindle-shaped form (**B**), *Inaria* (**C**), *Cyclomedusa* (**D**), and an unnamed bloated being (**E**) that lived prostrate on the sea floor. Whatever they were, the organisms that made up the Ediacaran biota were unique. All are extinct. By about 530 million years ago they had been replaced by shelled Cambrian animals.

Swamp

Phylum: Rhizopoda. Pictured here are three types of rhizopods drawn to relative scale: *Amoeba proteus* (**A**), shown engulfing a ciliate (**B**); *Arcella polypora* (**C**), which lives in a hardened shell, or test; and *Mayorella penardi* (**D**), a polypodial ameba. Commonly called amebas, the rhizopods are single-celled, uninucleate, heterotrophic organisms that move by means of pseudopodia. The ameba protrudes a temporary foot (pseudopod) of cytoplasm, which is used for locomotion or for feeding. The moving, feeding form is called the trophozoite; the dormant stage (if there is one) is referred to as a cyst.

There are about two hundred species of ameba, including *Entamoeba histolytica,* the symbiont whose presence in large numbers is correlated with intestinal gas and diarrhea in humans. As a cyst, this microbe can tolerate the acidic gastric fluid that it encounters when entering the human intestinal lining. Its growth apparently causes symptoms of amebic dysentery. Ameba cysts are found in drinking water and on fruits and vegetables washed in water contaminated with human feces.

Amebas like those pictured here eat phototrophic organisms such as cyanobacteria (see page 34) and algae (e.g., chlorophytes, see page 84). Other rhizopods eat heterotrophic bacteria and protists. All amebas, however, ingest their food by an engulfing process called phagocytosis. Many consume other protoctists (including other amebas), fungi, and bacteria. A mixture of amebas can be found in concentrations of 10^4/g to 10^5/g

of dry weight in sandy loam soil. Some amebas—like the uninucleate ones shown here (**E**)—are tiny, as small as 1 or 2 micrometers. In comparison, others are giant—up to 1,000 micrometers (1 centimeter).

Phylum: Karyoblastea. In the mud at the bottom of stagnant streams and ponds in temperate regions of the northern hemisphere is the curious unique microbe *Pelomyxa palustris* (**F**), sole species of the phylum Karyoblastea. Individuals often move slowly, looking like tiny drops of water, over the surface of decaying leaves and other brownish vegetation. Because it can grow to 5 mm long, this organism can be seen by the observant naturalist's naked eye.

This giant ameba differs from other amebas, such as rhizopods (see this page), in many ways. It is multinucleate but lacks mitochondria, Golgi bodies, and other organelles usually found in eukaryotes. It harbors three types of endosymbiotic bacteria, at least one of which functions like mitochondria. The other two actually rid *Pelomyxa* cytoplasm of excess hydrogen, not combining it with oxygen as mitochondria do, but rather as methanogenic bacteria do (see page 22): The hydrogen from food is combined with carbon dioxide, and methane is given off as waste.

Harboring two kinds of methanogenic symbionts allows *Pelomyxa* to live in microaerobic conditions. Although it does require some oxygen, *Pelomyxa* can survive lower concentrations of oxygen than exist in the atmosphere. Like all other amebas, karyoblasteans lack a sex life that involves meiosis or fertilization. They reproduce to form numerous binucleate cells or a few multinucleate ones. They consume algae and plant debris and were dubbed "exceedingly gluttonous animals," by Joseph Leidy, who in 1879 was the first to describe them.

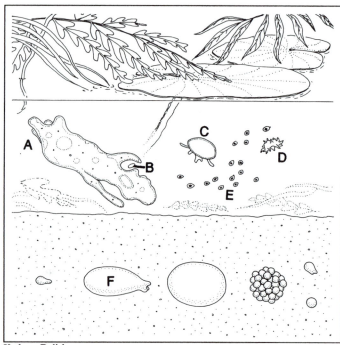

Kathryn Delisle

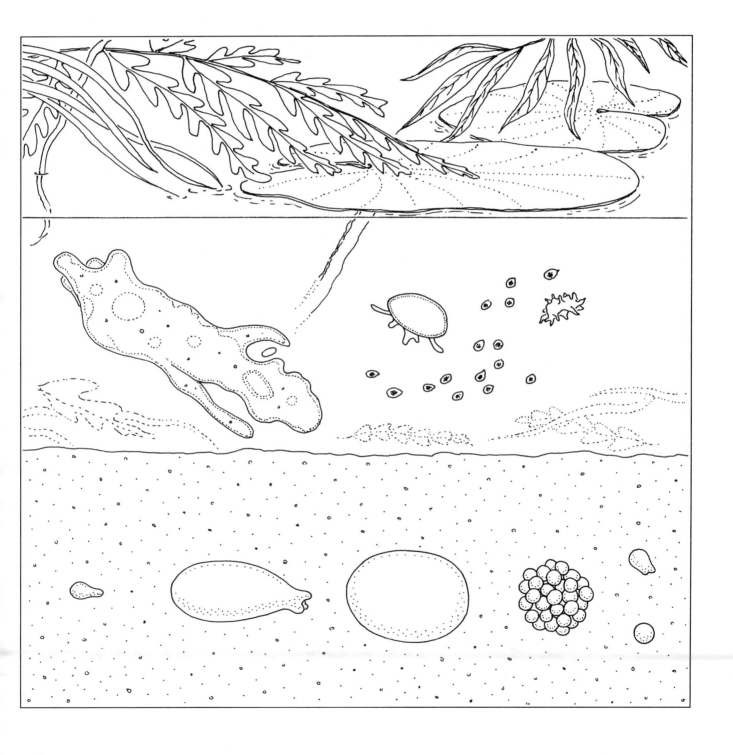

Estuary with Oyster Bar

Phylum: Haplosporidia. Haplosporidians are infamous because of their effect on the seafood industry. All haplosporidians live in the tissues of freshwater or marine animals. Some, such as *Haplosporidium nelsoni* (**A**), which has caused serious epidemics in the American oyster, *Crassostrea virginica* (**B**), a mollusc (see page 142), are destructive. As symbiotrophs, haplosporidians may be benign or pathogenic, a state that is believed to be a function of salinity and probably temperature. They are not found north of Massachusetts or south of Virginia, and most species can survive only symbiotically in mollusc tissue.

Unlike the superficially similar paramyxeans (see this page), haplosporidians typically form unicellular, uninucleate propagules, called spores. They live between the cells in animal tissue as uni- or multinucleate unwalled cells called plasmodia, and they contain haplosporosomes (**C**), membrane-bounded organelles of unknown function.

Phylum: Paramyxea. The paramyxeans, all of which live in the tissue of animals, are unique because of the way they form propagules: Through a process of internal cleavage, cells called spores are enclosed inside one another like Russian dolls. Because paramyxeans lack the attachment structures and apical complex, they are not apicomplexans (see page 98); because they lack haplosporosomes, they are not haplosporidians (see this page). Because of their internal cleavage—the formation of cells inside of cells—paramyxeans are given their own phylum.

The three genera and six species of paramyxeans are distinguished by the number of spore cells and by the taxon of the insect (see page 144) or mollusc (see page 142) in which they reside. The best-known paramyxeans are troublesome pathogens that infect commercially important seafood—bivalve molluscs such as oysters and clams. *Paramyxa paradoxa* grows necrotrophically in the larval intestinal epithelium of annelid worms (see page 136) such as *Poecilochaetus serpens*.

Before electron microscopy helped clarify the distinctions between apicomplexans (see page 98), haplosporidians (see this page), microsporans (see page 58), and paramyxeans, all of these organisms were jumbled together and maligned as "parasites." Shown here is *Marteilia refringens* (**D**), a multicellular paramyxean with a small number of cells (fewer than twelve) per individual that lives in large numbers in the European oyster, *Ostrea edulis*.

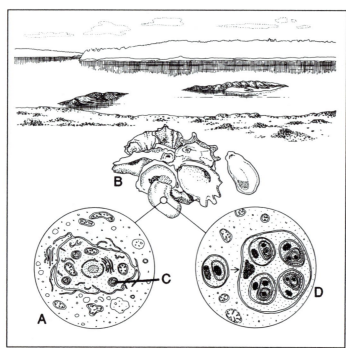

Kathryn Delisle

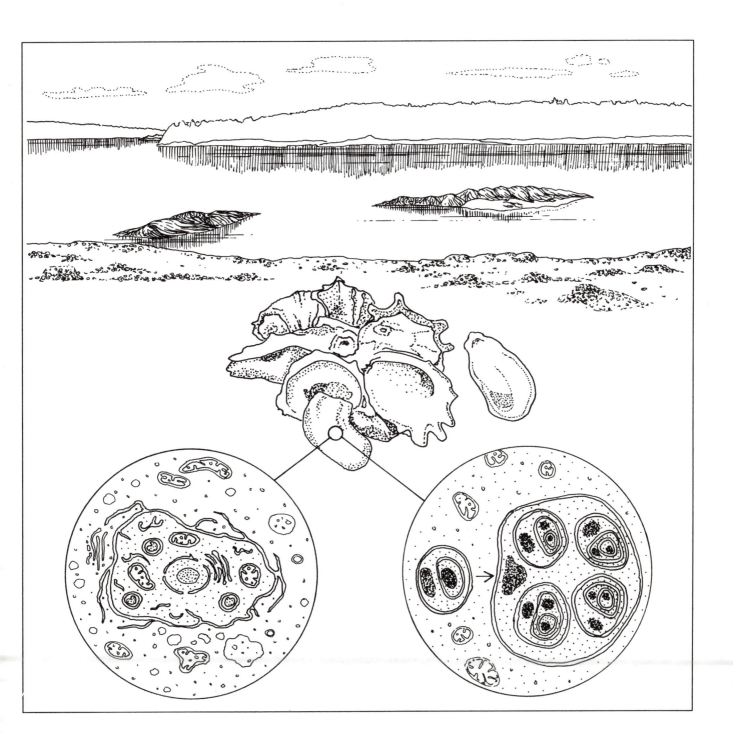

Salmon

Phylum: Myxozoa. The myxozoans once were thought to belong to a broad grouping of protoctists called sporozoans. Unlike other former sporozoans—now recognized to be apicomplexans (see page 98), microsporans (see page 58), or paramyxeans (see page 54)—which are single-celled protists, myxozoans are multicellular. Each propagule has two nematocyst-like structures called polar capsules with filaments inside as shown here (**A**). These polar capsules release the filaments that, like a dog's leash, attach the myxozoan to the intestinal tissue of the animal in which it resides.

The phylum is divided into two classes: the Myxosporea, with over a thousand species, and the Actinosporea, with only thirty-seven species. Included in the phylum are important symbionts of fish, some of which cause serious diseases. *Myxobolus cerebralis,* for example, lives in sockeye salmon (*Oncorhynchus nerka,* **B**). Pictured here are the infection stages within the salmon. The sticky polar filaments are extruded (**A**), and the sporoplasm is released (**C**) with haploid nuclei that later fuse in a sexual event of fertilization, which occurs before development of the multinucleate plasmodium.

The plasmodium (**D**) has two types of nuclei: haploid nuclei with one set of chromosomes and diploid generative cells that form after karyogamy. A cross section (**E**) of a mature myxozoan propagule (still referred to by many as a spore) shows the two polar filaments inside the capsule.

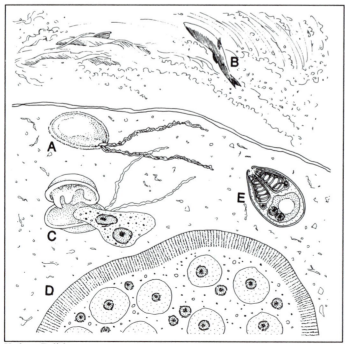

Kathryn Delisle

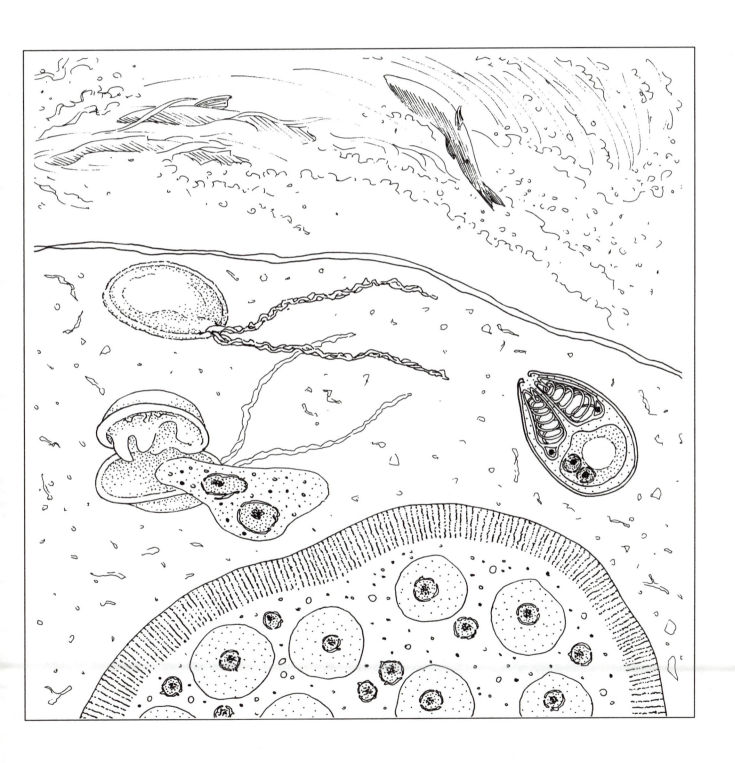

Continental Shelf with Flounder

Phylum: Microspora. All microsporans (also called microsporidians) are small, unicellular symbiotrophs that must live inside the cells of animal tissues. Unlike most protists, these organisms lack mitochondria at all stages of their life history; microsporans may even be relics of protist life before mitochondria evolved. Nearly all arthropods (see page 144) and chordates (see page 142) harbor microsporans unique to their species. Microsporans are found in all classes of vertebrates, especially fish. *Glugea stephani* (bottom), for example, lives in the starry flounder (*Platiclothus stellatus*, **A**).

Microsporan bodies are modified as unique infection devices, which have been called spores because they appear as small, mysterious, dark propagules and are usually present in great quantity. These "spores" form the infective structure, called the polar tube (**B**), which

is only 0.1 micrometer in diameter but is 100 micrometers long and is coiled up within the spore cell. When released, the tip of the tube jabs an animal cell, allowing the nucleus and cytoplasm of the microsporan cell to flow through the tube and inoculate the tissue cell of the animal. Once inside, the cell forms a multinucleate plasmodium (**C**), which then undergoes sporogony (that is, multiple fission) to form many offspring microsporans at one time.

In some cases the microsporan infection resembles a tumor because the infected animal cell survives, but it grows, accommodating more and more microsporans until it depletes the food supply from the surrounding tissue. The microsporan-induced, single-cell tumor (hypertrophied animal tissue cell) is such a unique structure that it has a special name: a xenoma.

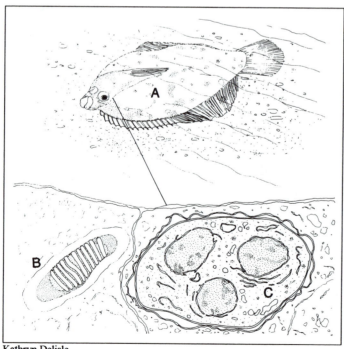

Kathryn Delisle

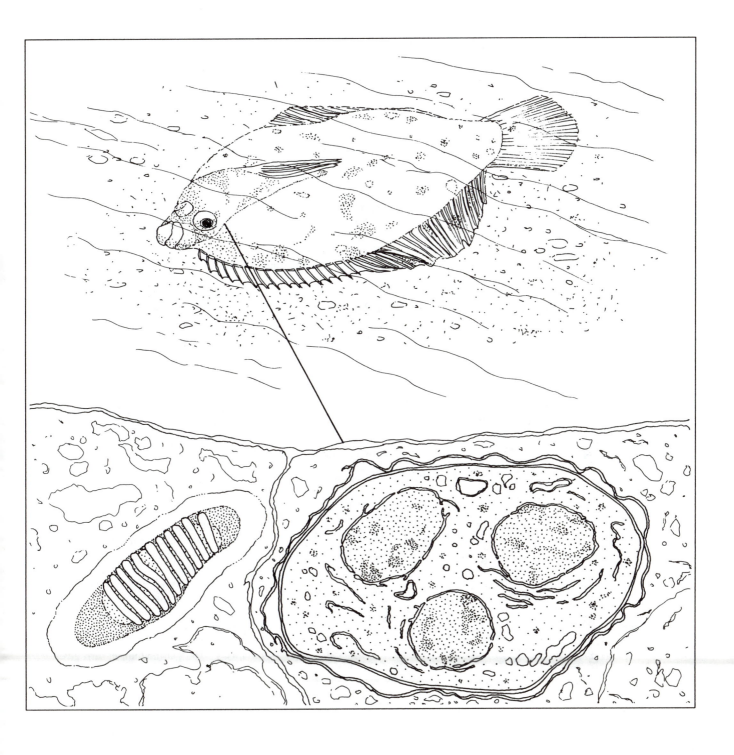

Birch Forest Floor

Phylum: Acrasea. The acrasids are cellular slime molds. They live freely as bacteria-eating, heterotrophic protoctists in fresh water and rotting vegetation. They abound as slime on fallen logs in woodland settings. Although meiotic sexuality is rare, cellular slime molds display an elaborate sort of parasexuality: orgies where hundreds of companions come together and fuse. Fusion leads to reproduction because propagules are formed that distribute the former amebas away from areas where they have exhausted the food supply.

Individual amebas of the cellular slime mold aggregate to form a structure called a pseudoplasmodium (because the ameba cells inside are still intact, so the structure is genuinely multicellular). The aggregate takes on the appearance of a slug. As conditions become drier, the slug differentiates into a sporophore with a sorus that contains dark, resistant propagules, which, except for being on the top of an erect sporophore, are virtually indistinguishable from ameboid cysts. These propagules are carried away by wind; some develop into amebas. Because of their unique differentiation from amebas to slugs to erect cyst-bearers, cellular slime molds are studied extensively by cell biologists. Two are shown here: *Copromyxella spicata* (**A**) and *Acrasis rosea* (**B**, ingesting bacteria).

Phylum: Dictyostelida. Dictyostelids are cellular slime molds with complex life-history stages: amebas, aggregation forms, and stalks that bear spores. Dictyoste-

lids constitute a phylum distinct from their closest relatives, the acrasids, because of several differences: Dictyostelids have better-differentiated stalk and spore cells, form more distinct sorocarps (the spore-bearing structures), and align into throbbing streams during ameba aggregation.

Because morphogenesis is separate from differentiation in its life cycle, *Dictyostelium discoideum* is used in laboratory experiments to study development. Dictyostelid spores germinate into amebas, the feeding stage of the cellular slime mold. As food or sunlight diminishes, the amebas aggregate into a visible grex. The aggregating ameba and shape-changing grex comprise the morphogenetic phase; the sporophore differentiates as basal disc, stalk, and spore cells.

Here amebas of different strains fuse to form a diploid zygote, called a giant cell, which then engulfs other amebas. As more amebas congregate near the giant cell, they become enclosed in a sheath and grow to form the macrocyst. Shown here (**C**) is the macrocyst of *Polysphondylium violaceum* with the giant cell in which meiosis occurs visible in the center. This structure forms a slug, which migrates. Later it develops into the spore-bearing sorocarp.

Phylum: Plasmodial Slime Molds. The plasmodial slime molds are denizens of the forest floor and its decaying trees. Unlike cellular slime molds, these organisms are composed of myriad nuclei not organized into cells; rather, the nuclei and other cell components move back and forth inside the multinucleate mass called the plasmodium. The movement is called intraplasmodial or cytoplasmic streaming.

Depicted here (**D**) is the large yellow myxomycote *Fuligo septica*. Its sporophores (**E**) are 1 to 5 cm in diameter and can be seen on well-fertilized lawns and logs. The sporophore consists of a basal disc, a stalk, and a sporangium or sporocarp, the stage in the myxomycote life history that produces propagules. Diploid spores produce either haploid undulipodiated swarmer cells (**F**) or haploid amebas, superficially like free-living amebas, called "myxamebas." Unlike other amebas, myxamebas can fuse with similar cells to produce a zygote that grows by nuclear division without cytoplasmic division to form the multinucleate plasmodium (**G**). The plasmodium in turn gives rise to stalked sporophores (**E**).

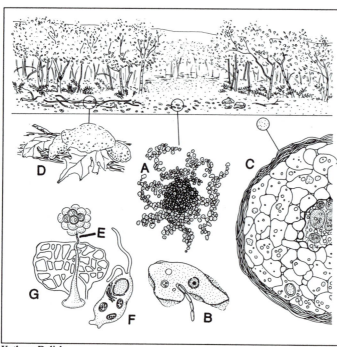

Kathryn Delisle

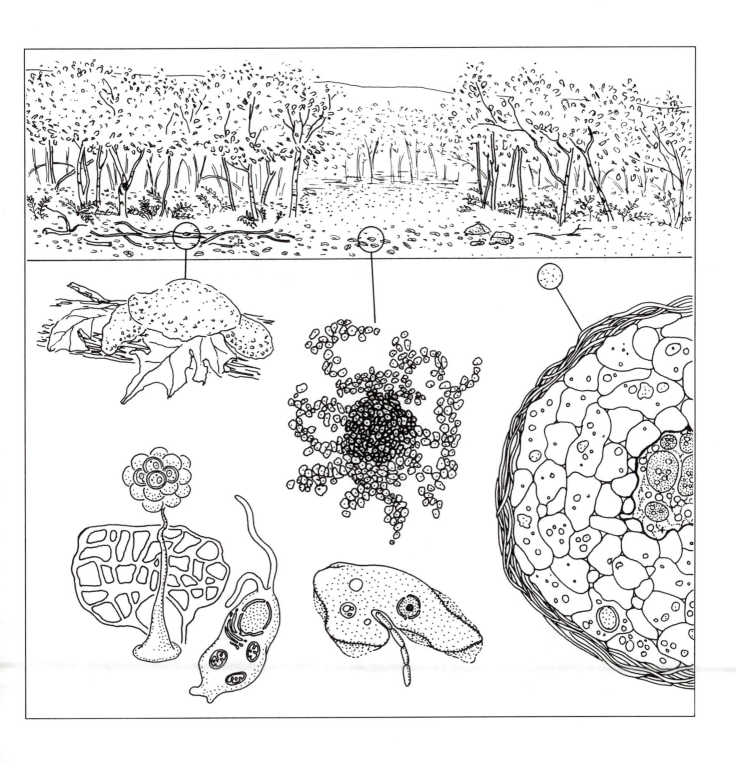

North Atlantic Coast

Phylum: Rhodophyta. Rhodophytes are red algae, a great marine group containing some of the largest multicellular protoctists, including many common seaweeds. Because they do not form diploid embryos retained in maternal tissue, rhodophytes are not plants. Red algae differ from aquatic plants in many other ways as well. The two classes of rhodophytes are the Bangiales and the Florideae, such as *Polysiphonia harveyi* (**A**). The cells of Florideae interconnect by small holes in their walls through which cytoplasm can flow. Called pit connections, the holes are visible in this detail of a cell (**B**).

No rhodophytes have undulipodia, but many have sexual stages. Small, spherical, immotile male cells, carpospores, are shed in large numbers over female threads, which they penetrate by growth. The life histories of some Florideae are very complex, whereas some Bangiales reproduce only by binary fission, showing no evidence of sex or gender.

All rhodophytes have reddish plastids (called rhodoplasts) that contain chlorophyll *a* and phycobiliproteins, which color the algae red. This phylum includes 4,000 species, including the coralline algae, which form calcium carbonate in their cell walls. Red algae are the material used in making agar, carrageenan, and other polysaccharides for the manufacture of ice cream and other food products. The agar gels used in DNA biotechnology could not be made without the products of *Gracilaria,* a prolific red alga. Some red algae are consumed whole, such as *Porphyra,* which is eaten in *gim bob* (Korea) and *sushi* (Japan).

Phylum: Phaeophyta. Phaeophytes, or brown algae, are the largest protoctists. Some forms, like the giant kelp, grow to up to a hundred meters long. Marine biologists speak of kelp forests, where thick populations bob with the waves. The Sargasso Sea is named for its great *Sargassum* alga. Phaeophytes live on rocky coasts around the world, dominating the intertidal zone and shoals near the shore. They are the primary producers for many communities of marine animals and microbes.

Phaeophyte sexuality is depicted here in the species *Fucus vesiculosus.* The diploid form, or sporophyte, is the conspicuous seaweed itself (**C**). On its surface (the thallus) the sporophyte produces little bumps, which are sex organs: antheridia (**D,** shown releasing male gametes) and oogonia (**E,** shown releasing female eggs). The tiny male gametes (**F**) are motile sperm cells, each with two undulipodia. One undulipodium is long and has mastigonemes; the other is short and whiplike. The sperm swim to the oogonium and fertilize the eggs. The fertilized egg then germinates, grows rhizoids, and becomes the new sporophyte. Brown algae, such as *Laminaria,* are eaten and used in dyes and adhesives. Phaeophytes are also a source of vitamins, medicines, and minerals.

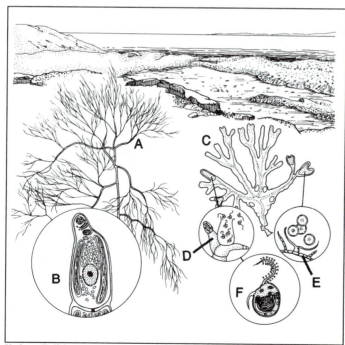

Kathryn Delisle

Shallow Pond

Phylum: Conjugaphyta. This phylum of freshwater green algae, called gamophytes, is distinguished from all other green algae by its sex life. Gamophytes conjugate: During the life cycle two bodies of more or less equal form come together to mate and fuse nuclei in the absence of any sperm or eggs. No undulipodia are ever formed by gamophytes. The group contains some of the most beautiful aquatic microorganisms known, although to the unaided eye and the unenlightened observer they appear mostly as floating pond scum (**A**). Many are green filaments; the others, such as *Micrasterias* (**B**) and *Cosmarium* (**C–F**), belong to a group that look like little jewels, called desmids.

Zygnema sp. (**G**) forms an algal mass. Individual cells of *Zygnema* (**H**) show the nuclei and stellar (star-shaped) chloroplasts with pyrenoids. The stellar chloroplasts are also visible in these two conjugating strands of *Zygnema* (**J**). *Micrasterias denticulata* reproduces asexually; shown here are two halves budding (**B**). Strands of *Spirogyra* (**K**) are recognizable by their characteristic helically wound chloroplasts.

Also shown are free-floating individuals of the desmid *Cosmarium* sp. (**C**). A single cell of *Cosmarium* (**D**) shows the nucleus and chloroplasts. In desmid sexuality the spiny ameboid zygotes leave their tests (shells) to conjugate (**E**). The resulting spherical resistant body (**F**) falls to the bottom of the pond until spring, when it returns to an active growing stage.

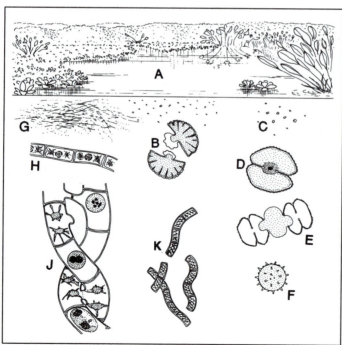

Kathryn Delisle

Marine Abyss

Phylum: Xenophyophora. From the most remote depths of the ocean, the abyss, come the rarely seen, strange protoctists known as xenophyophores (in Greek *xenos* means "foreign," and *phyo-* means "to produce"). These little-known heterotrophic organisms often have been mistaken for fecal pellets, mineral deposits, or foraminiferan tests because they cover themselves by cementing pieces of foreign material together.

The traditional, rough sampling techniques used in investigating the abyss usually break these protoctists into pieces, increasing the confusion that surrounds them. Better understanding of the deep-sea vents, cold seeps, and other features of the sea floor, as well as gentler methods of sample collecting, now provide an abundant supply of xenophyophores for study. Thirteen genera and forty-two species have been described, but by only a few scientists.

The xenophyophores are plasmodial organisms—i.e., they are formed of a multinucleate common cytoplasm that is inside a cemented, branched tube system called a granellare. The cytoplasm of many xenophyophores contains crystals of barite (barium sulfate). Several species are depicted here: *Galatheammina tetradea* (**A**); *Reticulammina lamellata* (**B**, with its pseudopods extended for phagocytotic feeding); the flat *Stannophyllum zonarium* (**C**, on the ocean bottom); and *Psammetta globosa* (**D**).

Stannophyllum can grow very long for a protoctist (up to 23 centimeters), but even at this length it remains only 1 millimeter thick. Almost no one has seen these protoctists alive. Fossils (e.g., *Paleodictyon*) in sedimentary rocks look much like xenophyophores; whether the modern and ancient beings are related, however, is still unknown. (See page 146 for identification of **E**.)

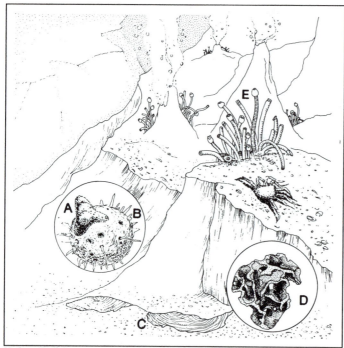

Kathryn Delisle

Ice-covered Lake

Phylum: Cryptophyta. Also known as cryptoprotists or phytoflagellates, cryptomonads are ubiquitous and abundant photosynthetic, motile, aquatic protoctists. Some form colonies of mucilaginous sheaths. Cryptomonads are a major constituent of the microbiota in ponds in early spring. Like euglenids (see page 74), these organisms photosynthesize but also may eat. If cryptomonads have any sexual life, it has gone undetected. They are generally found in cold environments. The reddish brown color of the red-tide ciliate (see page 94), *Mesodinium rubrum*, in the ocean is due to cryptomonad endosymbionts that are partly photosynthetic. Hundreds of these photosynthetic cryptomonad symbionts reside in the cytoplasm of just one of these fast-moving ciliates.

The cryptomonads, such as *Cyathomonas truncata* (**A**), differ from euglenids in that they have two anteriorly directed undulipodia of unequal length inserted in a gullet, or crypt. This cross section of *Cyathomonas* (**B**) shows its organelles. Cryptomonads reproduce by a distinctive form of cell division: A second crypt forms (**C**), one of the two crypts migrates to the opposite end of the cell (**D**), and the cell divides (**E**). Many also have trichocysts, which are expelled to sting and capture small protists or bacteria. About sixty species are known; some, like *Copromonas*, eat only bacteria, lack photosynthesis, and grow well in the dark in the laboratory. In nature, they dwell in dung.

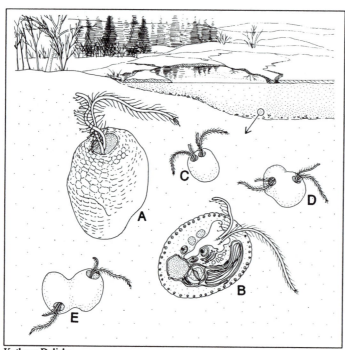

Kathryn Delisle

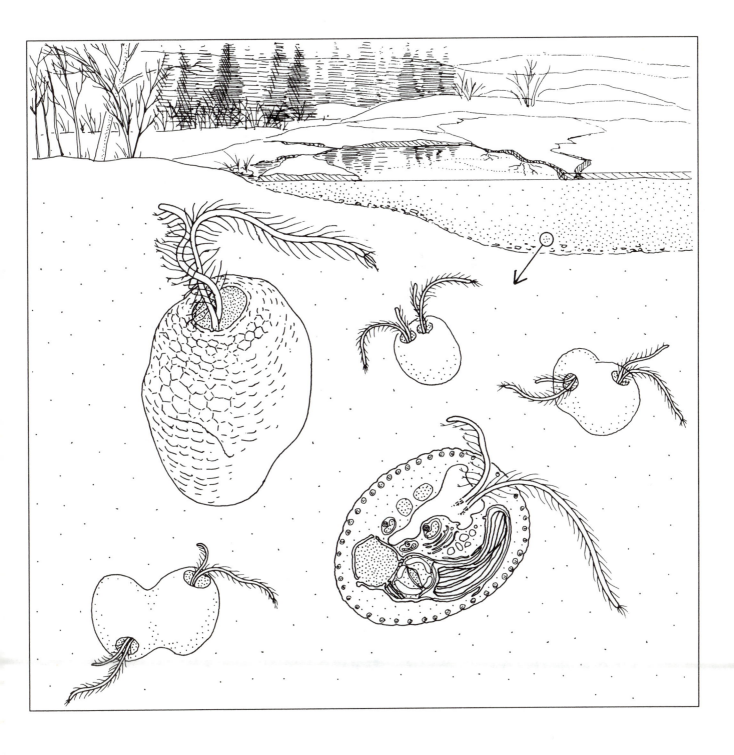

River Delta

Phylum: Glaucocystophyta. Glaucocystophytes are rare freshwater algae that live photoautotrophically (producing their own carbon products by using sunlight) with the aid of bright bluish green, chloroplast-like bacterial symbionts called cyanelles. Lacking a cell wall, these symbionts are surrounded by membranes: They lie in a vesicle. Glaucocystophytes are either mastigote or coccoid algae. The motile forms have two undulipodia, both with brushlike mastigonemes. *Gloeochaete* sp. (**A**) produces stiff, hairlike extensions called pseudocilia arising from the apical depression (**B**).

Nine genera and thirteen species of glaucocystophytes are recognized. Although they have limited biochemical and genetic makeup, the cyanelles inside the cells of some species of glaucocystophytes, such as *Cyanophora paradoxa* and *Glaucocystis nostochinearum*, have been classified as symbiotic cyanobacteria.

Phylum: Chlorarachnida. Although others are thought to exist, to date only one chlorarachnid—*Chlorarachnion reptans,* a phototrophic ameba that lives in a plasmodial reticulum network (**C**)—has been formally described. Each cell has several bright green plastids (chloroplasts), making the ameba capable of photosynthesis, although it eats protoctists phagotrophically. *Chlorarachnion* has a distinctive periplastidial compartment in the region near the base of the pyrenoid, where starch is formed (**D**). Distinctive zoospores (**E**) form and disperse *Chlorarachnion* cells through the sea water

in which they live. A single undulipodium is wrapped helically around the body of the cell as it swims (**F**).

Once considered a xanthophyte (see page 78) or a foraminiferan rhizopod (see page 96), *Chlorarachnion* (whose name means "green spider web") is an extremely atypical alga. Its plastids have chlorophyll *b,* and its mitochondria have tubular cristae (the folded membranes), whereas other algae with chlorophyll *b* have mitochondria with flat cristae.

Phylum: Raphidophyta. Raphidophytes are tiny, obscure algae that, even to specialists—phycologists and marine biologists—are not well known. Detailed structural analysis is required for proper identification of the genera. Raphidophytes are unicellular, wall-less algae that live as plankton do, among aquatic plants or adjacent to the mud. They are heterokont (i.e., they have two different undulipodia), but only the forwardly directed undulipodium bears the tubular hairs called mastigonemes.

At least one species, *Olisthodiscus luteus,* is quite prolific. Scientists have determined that the dominant planktonic organism in the Narragansett Bay (Rhode Island) in summer is this raphidophyte, which has a large Golgi body (or dictyosome), used in secretions, adjacent to the central nucleus.

Only four genera and nine species of raphidophytes have been described. Shown here are *Vacuolaria* sp. (**G,** growing among water weeds) and the marine species *Chattonella* sp. (**H,** in cross section), which was named after the French marine biologist Edouard Chatton (1883-1947), who was the first to recognize the difference between prokaryotic and eukaryotic microorganisms. Chatton lived at the marine station Laboratoire Aragon on the western French Mediterranean and worked tirelessly from approximately 1920 until his death in 1947 to make known the complex life histories of marine protoctists.

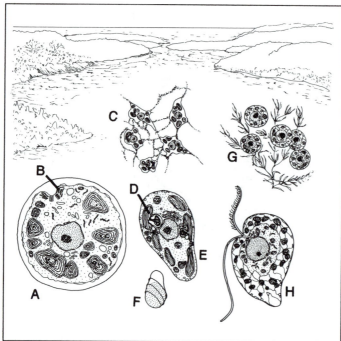

Kathryn Delisle

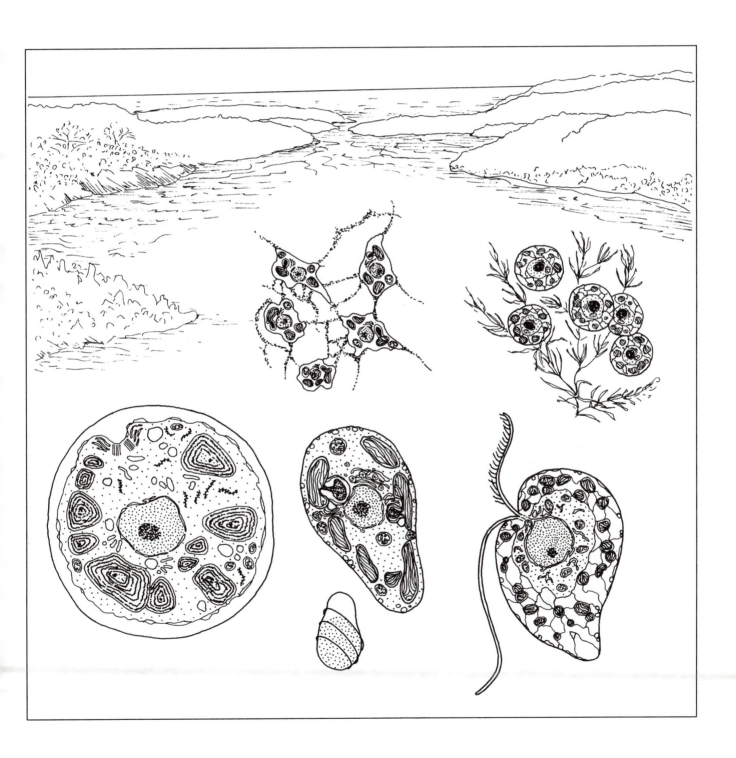

Fallen Log

Phylum: Zoomastigina. The zoomastiginids are a miscellany of small heterotrophs, most of whom derive their food osmotically. Some zoomastiginids interact with people as necrotrophs and therefore have enjoyed much more scrutiny than the rest. Zoomastiginids in general have suffered a bad reputation because of a few that are associated with human diseases: *Leishmania* (leishmaniasis), *Giardia* (giardiasis), and *Trypanosoma* (African sleeping sickness).

Although more will be identified as they become better known, at present eleven classes of zoomastiginids are recognized. Amebomastigotes are free-living and transform between the ameba pseudopod and mastigote swimming stage. Many form resistant cysts, like *Naegleria*. Bicosoecids and choanomastigotes (traditionally called choanoflagellates) are also free-living and inhabit marine waters. Diplomonads have double sets of nuclei; they attach to the intestines of mammals, thereby sharing their food. Kinetoplastids are distinctive because they have an organelle related to a mitochondrion called the kinetoplast. (Trypanosomes are kinetoplastids.) Opalinids are large for protists; they swim around in the cloacas of frogs and toads. Parabasalians and pyrsonymphids tend to live in insects. Proteromonads live inside animals; all proteromonads lack mitochondria, whereas pseudociliates, like *Stephanopogon*, have mitochondria and live free in water. Retortamonads are symbiotrophs in vertebrates.

Most zoomastiginids are unicellular with at least one undulipodium; some have thousands of undulipodia. Three representative genera of parabasalians are shown here: *Lophomonas* (**A**), *Trichonympha* (**B**), and *Trichomonas* (**C**). None have mitochondria, but many harbor bacterial symbionts.

In this scene of termites is a hidden component. Dry-wood–eating termites, *Pterotermes occidentis* (**D**), eat and nest in the logs of *Cercidium* (**E**)—paloverde, a leguminous tree—but it is the microorganisms in the gut of the termite that digest the wood. The termite itself ingests but cannot digest any wood. In this symbiosis, the parabasalian protists digest the cellulose from the wood into sugars for themselves and form acetate, which enters the cells of the termite intestines, thus nourishing these colonial insects.

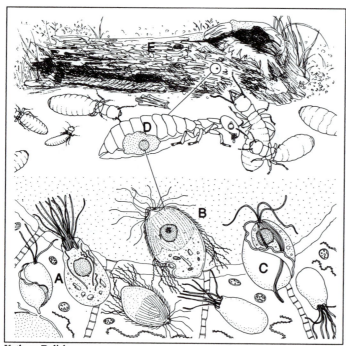

Kathryn Delisle

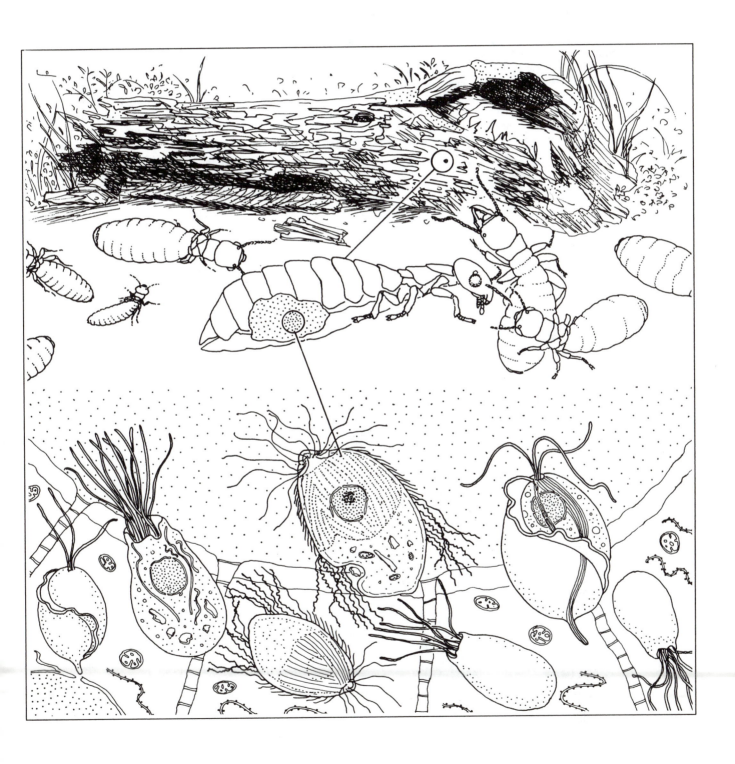

Outfall Pipe

Phylum: Euglenida. Because they exhibit the classic characteristics of both animals and plants, euglenids are the perfect example of why the two-kingdom classification is archaic. Like animals, euglenids are motile because they have a single, anteriorly directed undulipodium. Like plants, they are packed with chloroplasts for photosynthesis. Euglenids differ from green algae (chlorophytes, see page 84) because they have a different combination of colored compounds in their plastids and their cytoplasm, and they lack a cellulosic cell wall. Instead they have a proteinaceous pellicle—a ridged, flexible, highly characteristic cell covering.

Euglenids differ from the conjugating green algae (see page 64) because they have no sex. Of the approximately eight hundred species of euglenids that have been reported in the literature—even those that are colonial, multicellular forms—not one has been seen engaging in sexual activity. All euglenids reproduce by binary fission, starting at the undulipodiated end of the cell. Some species, such as *Euglena gracilis,* have been useful for analyzing cell organelles. For example, the cell can lose its chloroplasts and, if well fed, survives and reproduces without them. *Euglena gracilis* has been used as an indicator in water-quality studies and in bioassays for vitamin B_{12}.

Pictured here are *Euglena spirogyra* (**A**) and *Phacus* sp. with its ribbed pellicle (**B,** external and internal views). Two stages of division during the reproduction of *Euglena gracilis* are shown in longitudinal cross section (**C**).

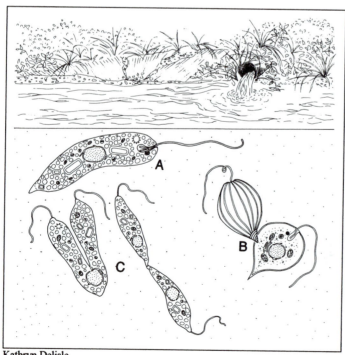

Kathryn Delisle

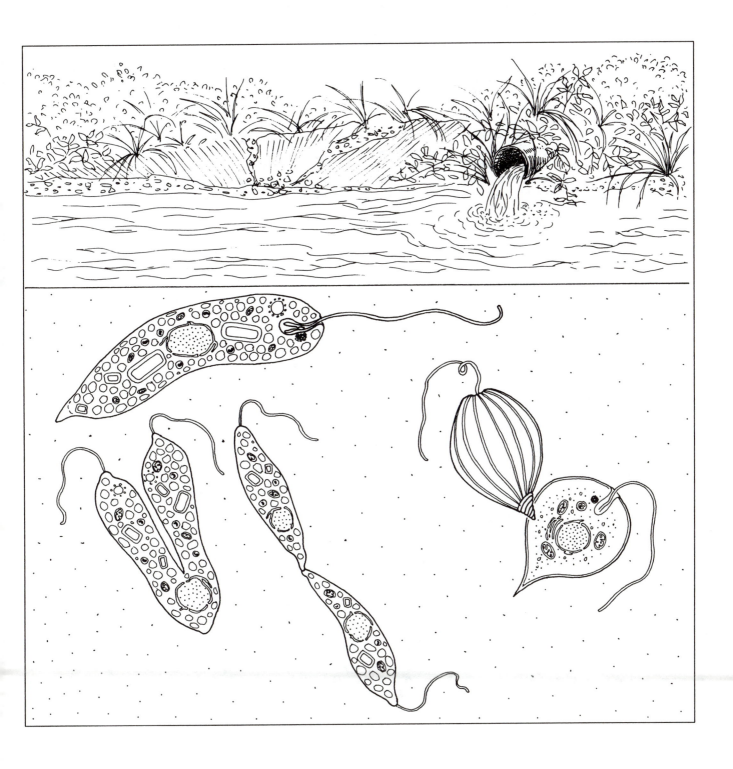

Marine Chalk Cliffs

Phylum: Prymnesiophyta. Organisms that float with the waves are called plankton by ecologists. If capable of photosynthesis, they are called phytoplankton. Since *phyto-* means "plant," however, this term should be replaced by "photoplankton." Prymnesiophytes (also called haptophytes or haptomonads) are typical of "cosmopolitan" photoplankton. Mostly single-celled algae, prymnesiophytes have chrysoplasts, just as the chrysophytes do (see page 90), but are mainly marine.

Some prymnesiophytes also form the distinctive resting stage called the coccolithophorid, a single cell covered with little calcium carbonate buttons or plates called coccoliths, which means "berry stones." Fossilized coccolithophorids and forams (shelled granuloreticulosans; see page 96) make up the famous chalk cliffs of England. The prymnesiophyte *Phaeocystis poucheti* contributes a significant amount of the gas dimethyl sulfide to the atmosphere as a byproduct of manufacturing a substance that maintains its salt balance. (Compounds made by cells that regulate internal salt concentrations are called osmolytes.)

For many years scientists were not aware that the free-swimming stage of prymnesiophytes (e.g., *Prymnesium parvum,* **A**) and the coccolithophorid (such as *Emiliania huxleyi,* **B**) are two different stages of the same group of organisms. Coccoliths are distinguished by their shapes; some examples of different types of coccoliths include discoasters (**C**), rhabdoliths (**D**), pentaliths (**E**), and helicoid placoliths (**F**). No one knows what these scales do, but they may act as a sort of venetian blind that modulates the sunlight reaching the plastids.

How do these tiny algae construct their elegantly patterned covering? Within the motile stage, organic scales form inside an organelle called the Golgi apparatus. Calcium carbonate crystallizes on the scales in species-specific patterns. As these microscopic coccoliths assemble, microtubules move them out to the surface of the algal cell.

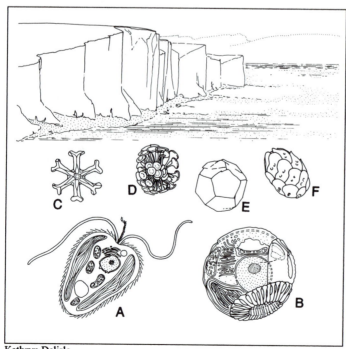

Kathryn Delisle

Lake Surface

Phylum: Eustigmatophyta. Eustigmatophytes are obscure, tiny, "eyespot" algae. They are yellowish green and contain a single, long xanthoplast like that of the xanthophytes (see this page). Their cell organization is completely different from that of the yellow-green xanthophytes, however.

The name Eustigmatophyta is derived from the presence in all of these organisms of eyespots with stigmas. Informally called eustigs, these inconspicuous protists tend to be known only to professional botanists and pond-water enthusiasts.

Generally found in fresh water, eustigs such as the representative genus *Vischeria* sp. form immotile, coccoid growing cells (**A**). Not known to engage in any type of sexuality, they reproduce by binary fission. The zoospore of *Vischeria* sp. (**B**) has a single, mastigonemate undulipodium with an undulipodial swelling at the base. An adjacent swelling, filled with drops of carotenoids, forms the eyespot (**C**). This organization of the cell probably assures that undulipodial movement directs the cell to optimally lighted environments, since the plastid absorbs visible light, which excites the carotenoid pigment.

Phylum: Xanthophyta. People tend to think of xanthophytes, often called yellow-green algae, as pond scum. Most xanthophytes are freshwater algae with photosynthetic pigments, including xanthins, which make them yellow-green. Many xanthophytes have been seen only rarely, probably because they exist in small numbers. Pictured are the colonial genus *Ophiocytium* (**D**) and *Botrydiopsis* (**E**), which resembles a bunch of grapes.

Xanthophytes overwinter as cysts. In the spring they germinate, changing shape as they become motile zoospores (**F**). The zoospore, a spermlike cell that swims but is capable of reproducing without fertilizing an egg, loses its undulipodia and begins to grow into the alga itself. Xanthophyte zoospores are heterokont: One undulipodium is longer than and different from the other: It is covered with mastigonemes. Most xanthophytes have yellow-green, disc-shaped plastids (photosynthetic organelles) that are situated against the cell wall.

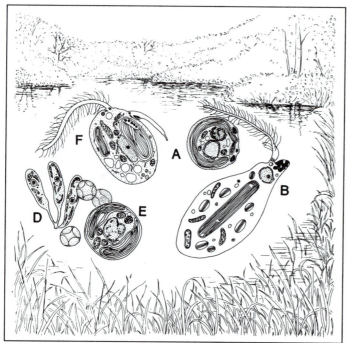

Kathryn Delisle

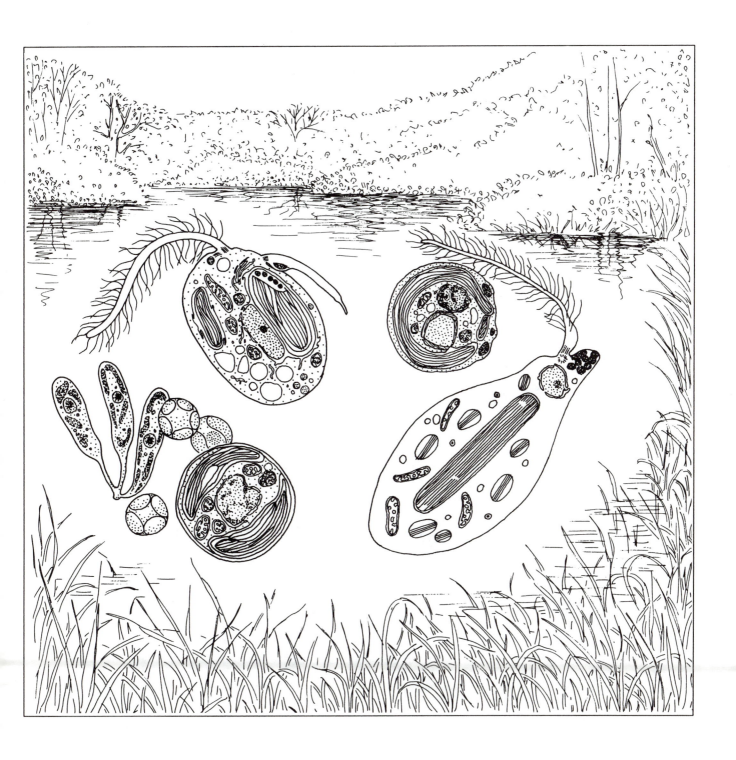

Ocean Water Column

Phylum: Actinopoda. Known to microscopists for three centuries as sun animalcules because of the rays that project from their cells, actinopods are distinctive, heterotrophic protoctists. Microtubules and filaments called axopods underlie their long, slender, projecting cytoplasmic organelles. In the plankton literature they are called radiolarians (in marine environments, **A**) or heliozoans (in fresh water, **B**), but ultrastructural studies have revealed their relationship to other protoctists.

The particular arrangement of the microtubules along the axis of the ray (called the axoneme, which means "axis thread") is what classifies the actinopods. Four groups are recognized: (1) Heliozoa, such as *Clathrulina fragilis* (**C**), shown here attached to a submerged leaf in a stream; (2) Acantharia from the sea, such as *Phyllostaurus siculus* (**D**), shown here with a cutaway section that reveals the spicule, cytoplasm-coated axopods (which are used to capture food), and capsular wall made of microfilaments; (3) Polycystina,

such as *Spongosphaera polyacantha* (**E**), shown here with a cutaway section that reveals the silica skeletal spine, axopods, and plates of the capsular wall; and (4) Phaeodaria, including *Challengeron wyvillei* (**F**), shown here with a cutaway section that reveals the endoplasm, nucleus, and a ball of waste product called the phaeodium.

Ameboid cells of phaeodarians are contained within plasmodial spheres, from which they are released as undulipodiated swarmer cells (**G**). These aquatic propagules develop without any sexuality and metamorphose into the adult. Polycystines, phaeodarians, and acantharians traditionally have been called radiolarians because of their marine habitat. Only electron microscopy and chemical analysis reveal the great differences among these floaters. The acantharian skeleton, unlike that of the others, is made of barium sulfate. Heliozoans covered with siliceous scales are common in freshwater streams and ponds.

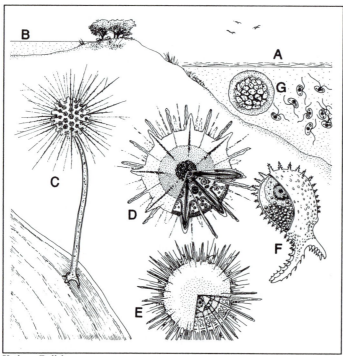

Kathryn Delisle

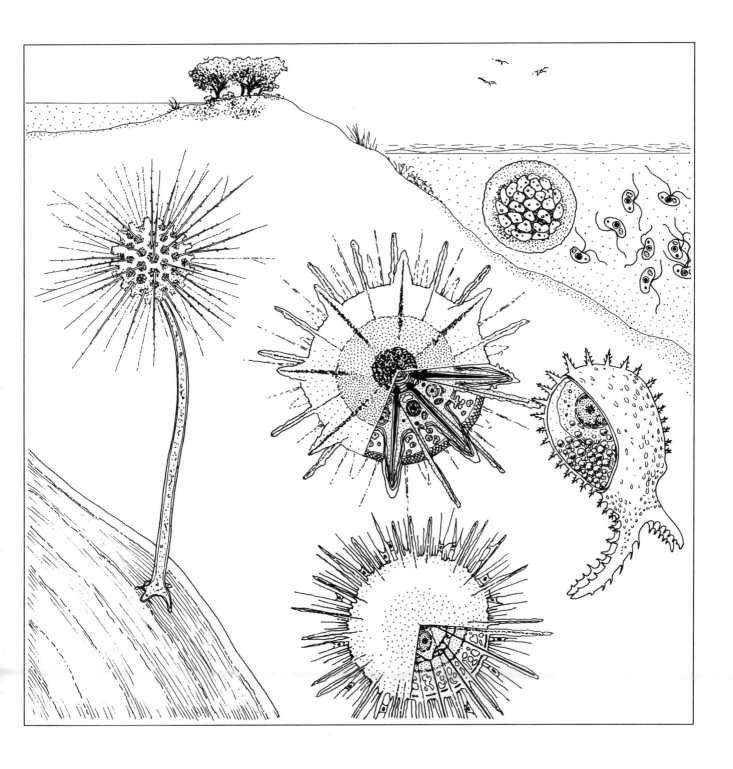

Pine Pollen

Phylum: Hyphochytriomycota. Because they look like white fuzz at a particular stage of their life history—just as plasmodiophorids (see page 86), chytridiomycotes (see page 92), and oomycotes (see page 102) do—hyphochytrids traditionally have been classified as fungi. They are more appropriately classified as protoctists, however, because they produce undulipodiated cells and lack the spores and life-cycle details of the asco- (see page 110) and basidiomycote (see page 112) fungi.

Hyphochytrids form swimming cells called zoospores that are capable of further development without fertilization. Each zoospore has a single, anteriorly directed undulipodium that bears mastigonemes. The spe-cies shown here, *Hyphochytrium catenoides,* grows on shed pine-pollen grains. The hyphochytrid cell produces a germ tube (**A**), which it uses to penetrate the cell wall of the shed pollen. Inside the pollen grain, the developing thallus (**B**), differentiating zoospores (**C**), sporangium (**D**), and protoplasm being discharged from the sporangium (**E**) of *Hyphochytrium* are visible. The magnified view of a hyphochytrid zoospore (**F**) shows the mastigonemes on its anterior undulipodium, as well as the nucleus, mitochondria, microtubules, and Golgi apparatus. To date, five genera and twenty-three species of hyphochytrids have been described in the professional (mostly mycological) literature.

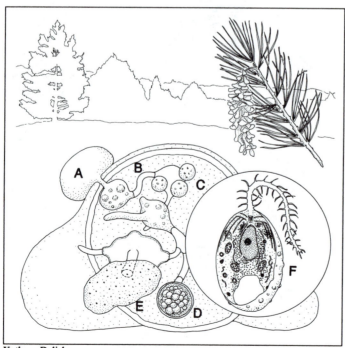

Kathryn Delisle

Atlantic Sheltered Bay

Phylum: Labyrinthulomycota. The labyrinthulomycotes, or slime nets, are colonial protoctists that form membrane-bounded ectoplasmic networks devoid of cytoplasmic constituents. Labyrinthulomycote cells produce these transparent networks from specialized organelles at the cell surface called sagenogens or bothrosomes. How cells within the colony move is unknown, but movement requires contact between the cell and the slime network and does not involve undulipodia.

Labyrinthulomycotes are heterotrophic, absorptive organisms that spend most of their lives moving back and forth inside the network, absorbing food osmotrophically. Some disperse via heterokont zoospores with two undulipodia. Labyrinthulomycotes are found mainly in shallow marine benthic environments associated with algae, with sea grasses such as *Zostera marina* (**A**), or with organic-rich sediments.

The representative genus, *Labyrinthula* (**B**), contains eight species. Careful studies of obscure marine protoctists called thraustochytrids unexpectedly revealed a relationship between them and the labyrinthulid slime nets. Now included in this phylum, therefore, are seven genera and thirty species of thraustochytrids. The cells of this arcane group of encysting protoctists also produce bothrosomes. Because they are clearly related to labyrinthulomycotes and they produce undulipodiated cells, thraustochytrids are no longer classified as fungi.

Phylum: Chlorophyta. Most green seaweeds and other green algae belong to this huge group of fascinating organisms, the chlorophytes. Many form zoospores, propagules that are capable of developing without a mate. Many also make very similar motile cells that are called gametes because, to develop, they must find a mate and fuse with it. Each zoospore or gamete has cup-shaped, grass-green chloroplasts and at least two anterior undulipodia of equal length. These are organisms of great translucent beauty and cosmopolitan distribution.

The 16,000 species of chlorophytes range in size from the 12-micrometer single cell of *Chlamydomonas,* invisible to the naked eye, to the conspicuous sea lettuce *Ulva lactuca* (**C**). The endobiotic chlorophyte *Chaetosiphon moniliformis* (**D**) grows in the decaying leaves of eelgrass (*Zostera marina,* **A**). Although many are marine, chlorophytes are a major component of the freshwater phytoplankton. Some live permanently on the bark of trees, others live in tree holes, and still others live chasmolithically—i.e., in the crevices of rocks. Chlorophytes are estimated to fix more than one billion tons of carbon in oceans and freshwater lakes each year.

Because of the structure of their walled cells and the way the cells divide, certain genera of chlorophytes, such as *Klebsormidium,* are believed to be the closest relatives of plants. Sexuality, followed by the formation of overwintering zygospores—hard-walled, resistant structures—is common in the phylum.

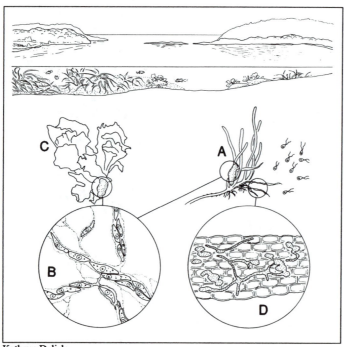

Kathryn Delisle

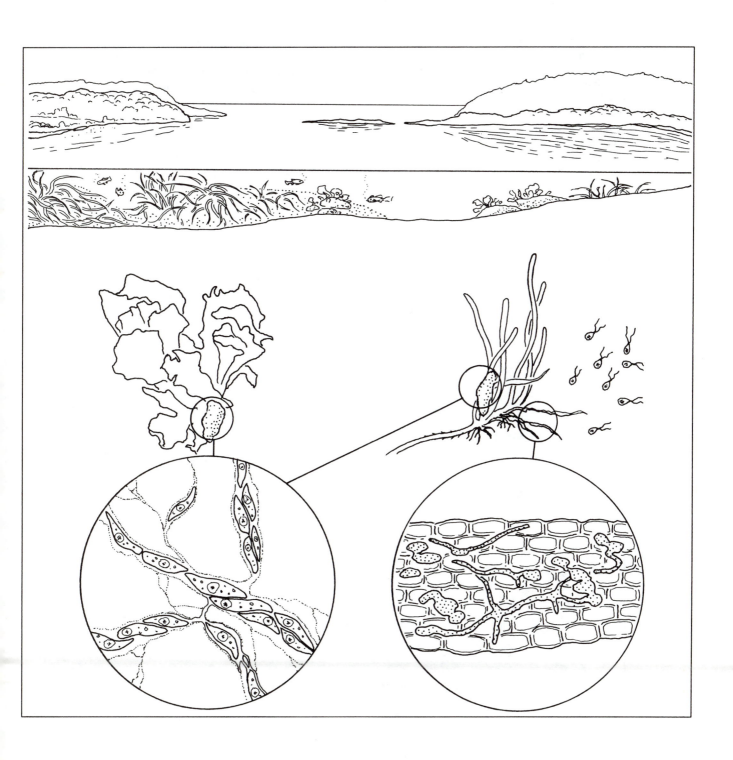

Cabbage Field

Phylum: Plasmodiophoromycota. The plasmodiophorids comprise another group of protoctists infamous for the disease-forming ability of a few of their members. Plasmodiophorids are obligate symbiotrophs; some grow on algae and fungi, whereas others grow on plants. The plasmodia, the multinucleate structures, are much smaller than those of the plasmodial slime molds (see page 60), and they live inside the cells of the organisms with which they are symbiotically associated.

Although the trophic stage is a multinucleate plasmodium, the life cycles and cell structures of plasmodiophorids are completely different from those of the plasmodial slime molds. For example, *Plasmodiophora brassicae* (**A**), which infects cabbage via the damp soil around the plant roots, requires two sets of zoospores and plasmodia to complete its life history. The infected cabbage plant (*Brassica oleracea,* **B**) appears much different from a healthy plant (**C**), whose fine root networks are free of infection.

This detail of a zoospore (**D**) of *P. brassicae* shows its two undulipodia—one directed forward, the other backward—as well as the nucleus, Golgi apparatus, microtubules, and mitochondria. The detail of the secondary sporangium (**E**) in a cabbage root hair shows the nucleus within the protoctist (denoted by the bold line), surrounded by the organelles (Golgi apparatus, mitochondria, and cytoplasm) of the plant cell.

Plasmodiophorids are studied primarily by scientists concerned about the destruction of important food plants—i.e., plant pathologists in schools of agriculture. Much more needs to be known about the details of the structure of the plasmodium, sexuality, cell behavior (the way the chromosomes clump and the nucleolus persists in mitosis), and many other aspects of the life of these obscure, often necrotrophic organisms.

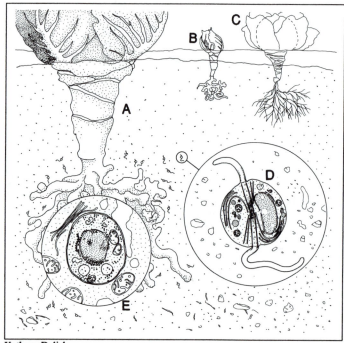

Kathryn Delisle

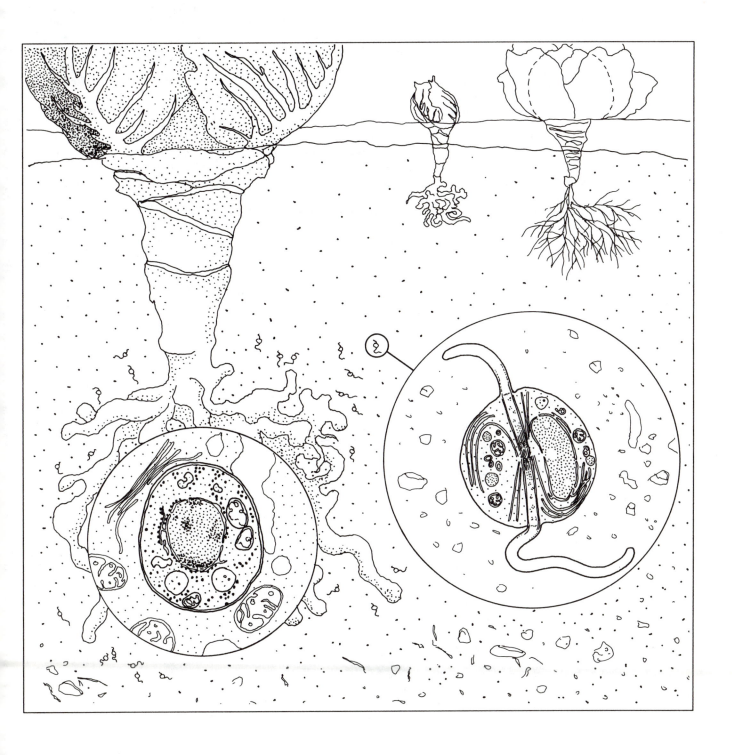

Coastal Red Tide

Phylum: Dinomastigota. Some people have estimated that there are 20,000 species of dinomastigotes. Even if you have never heard of them, you probably know what some of them can do. Dinomastigotes (also called dino-flagellates) are remarkably sculptured and diverse organisms. They are known mainly because some of them form the toxic red tides (**A**) that poison fish and shellfish. Some dinomastigotes are bioluminescent and cause a phosphorescent light on the sea at night. Others live inside the cells of coelenterates (see page 122) such as reef-forming corals and sea anemones. Most dino-mastigotes are covered with armored plates, which helped them fossilize well during the Phanerozoic eon.

The life history of the dinomastigote *Pyrocystis* has several stages, including the resting cyst (hypnocyst, **B**), the dormant spiny cyst (**C**), and the planozygote (**D**). This cross section (**E**) of the adult *Gonyaulax tamaren-sis* (**F**) shows its organelles, with a detail of the tricho-cyst (**G**), an organelle that discharges suddenly to sting prey. The arrangement of the two undulipodia—one longitudinal and one transverse—in a characteristic groove is distinctive for the entire phylum.

Dinomastigote DNA is referred to as mesokaryotic (literally, between prokaryotic and eukaryotic). Whereas chromatin is organized in fine fibrils like that of bacteria, it is clumped into chromosomes that can be seen (with a microscope) before, during, and after cell division. The complex coverings, or tests, of some dino-mastigotes have robust walls and permit the cells to overwinter in sediment. The tests bear exit pores through which the cells emerge in the spring. These cell features are unique to the 4,000 known species of dinomastigotes.

Typical cells of the colonial dinomastigote *Polykri-kos* pile one atop the other like Siamese twins. Some di-nomastigotes are necrotrophs on marine animals. Only detailed cell-structure studies, however, make their di-nomastigote affinities unequivocal. Fossil evidence dates dinomastigotes to at least the earliest times of the Cambrian period, 580 million years ago.

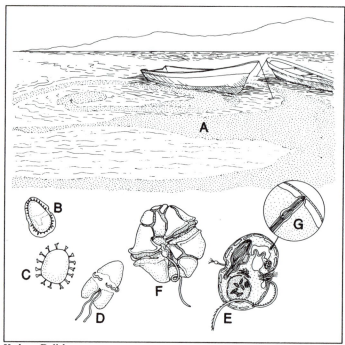

Kathryn Delisle

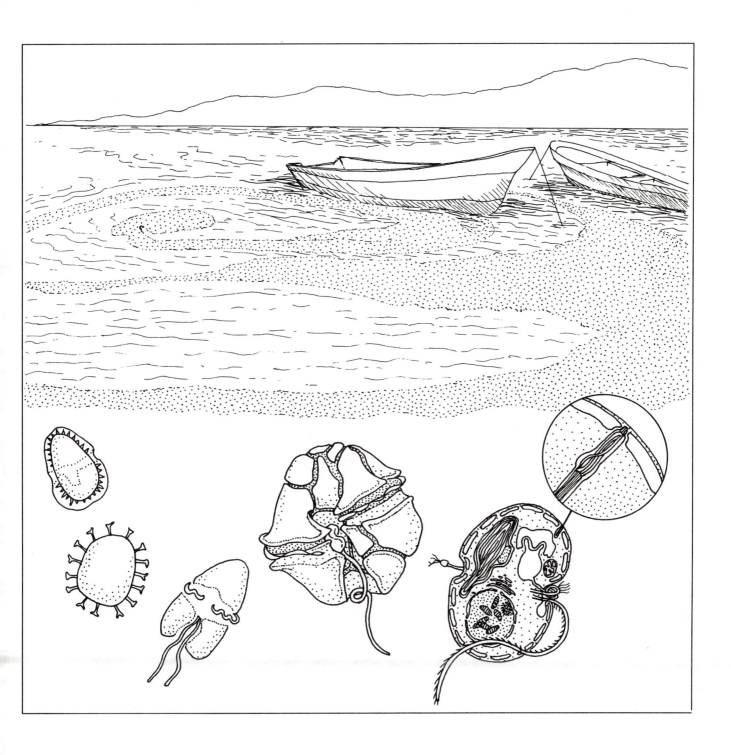

Farm Pond

Phylum: Chrysophyta. Also known as golden-yellow algae or chrysomonads, the chrysophytes are algae that look like brownish or yellowish scum to the naked eye. The group includes single-celled, individual plankton protists and larger forms. In the larger masses, the colonial forms, as well as with the single cells, each cell has chrysoplasts—organelles in which photosynthesis is based on chlorophylls *a* and *c*. Fucoxanthin is the most important accessory pigment. Chrysophyte swarmer cells or mastigotes are heterokont: Each has two undulipodia of unequal length, the longer of which has brush-like mastigonemes.

Chrysophytes are more common in fresh water, such as this farm pond (**A**), than in salt water. Yet one group, the silicomastigotes, which incorporate silica from sea water in their tests, is exclusively marine. Chrysophytes are useful to ecologists and geologists who reconstruct the past history of the environment. Many chrysophytes form beautiful sculptured silica scales that are resistant to decay. The scales of *Mallomonas* and other genera of Mallomonadaceae, for example, are entirely distinctive. They make useful paleoecological indicators of lakes because each species grows under specific conditions of temperature, light, phosphate concentration, organic content, and other ecological factors.

Shown here are the colonial chrysophytes *Synura* sp. (**B**, with a clump of cells breaking away to form a new colony) and the single-celled *Ochromonas* sp. (**C**), an organism used by cell biologists to study morphogenesis (the appearance of new form and other cell functions).

In the chrysophyte *Dinobryon* (**D**), typical chrysophyte cells are surrounded by vase-shaped loricas in a colony. These protoctists are relatively common and can be collected by students wishing to study the golden yellowish pond scum.

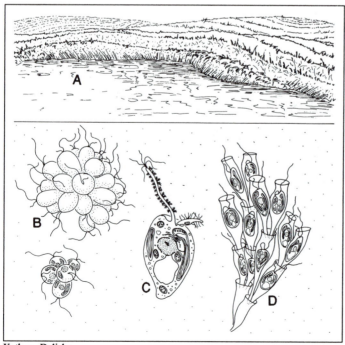

Kathryn Delisle

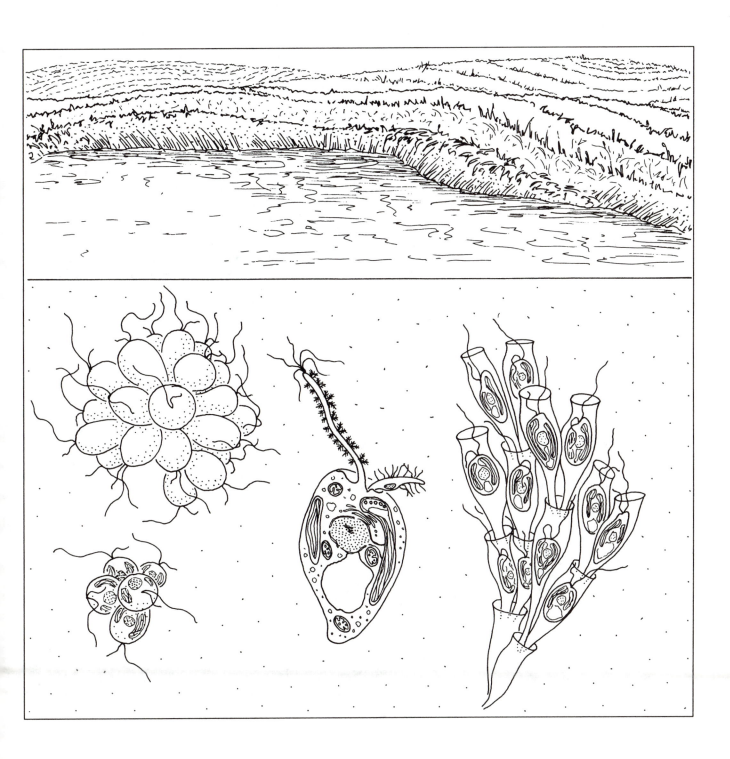

Decaying Pond Vegetation

Phylum: Chytridiomycota. Of all the funguslike protoctists, chytridiomycotes bear the closest resemblance to fungi. Some scientists argue that filamentous fungi were derived from these organisms by loss of undulipodia; others believe that chytridiomycotes are undeniably protoctists and that the earliest fungi were yeasts. Like fungi, chytridiomycotes contain a nitrogen-rich polymer of glucose called chitin in their cell walls and have absorptive nutrition. In most cases, they form a motile zoospore (which bears a single posteriorly directed undulipodium), as well as the chytrid body, a visible, cup-shaped structure in which the zoospores form.

Chytridiomycotes live on plant debris in lakes, such as decaying leaves, fruits, pollen, or seeds swept into the water. Some are symbiotrophic. The phylum contains a thousand species, divided into four orders: Chytridiales (chytrids in the strict sense), Blastocladiales, Monoblepharidales, and Spizellomycetales. Pictured here is *Rhizophydium granulosporum* (**A**) on the green alga *Oedogonium* (**B**), the chlorophyte (see page 84) from which *Rhizophydium* derives all its food. Here too is *Spizellomyces* sp. (**C**, with its discharge papillae), inhabiting a pine-pollen grain. The chytridiomycote eats out the plant material from inside. This detail of a generalized chytridiomycote zoospore (**D**) shows its organelles. Also shown is a close-up of the thallus of *Blastocladiella emersonii* (**E**), in which the rhizoids, basal cell, and apical sporangium, from which zoospores can be released in profusion, are visible.

At least five genera of chytridiomycotes—*Neocallimastix, Piromyces, Caecomyces, Orpinomyces,* and *Ruminomyces*—are obligate anaerobes. All of these grow inside the fermentation tanks (rumen or other modified digestive organs) of animals that have fibrous cellulose diets, such as cattle and elephants. Chytridiomycotes are thought to have lost their mitochondria and aerobic way of life after years of dwelling in the dark anoxic zones of animal intestines. Some of these protoctists engage in sex as spermlike swimming cells that find themselves attracted to female filaments.

Decaying pond vegetation provides a habitat for myriad microorganisms besides chytridiomycotes, such as algae, euglenids, filamentous bacteria, and many more. In the professional literature the habitat is called aufwuchs (from German) or periphyton (literally, "around the plants"). These aufwuchs or periphyton communities are home for many protoctists besides certain aerobic members of the Chytridiomycota. The soft, thick film of microorganisms that make up the periphyton community is easily observed on underwater stems of plants that grow in Florida's Everglades National Park and similar warm, sunny sites.

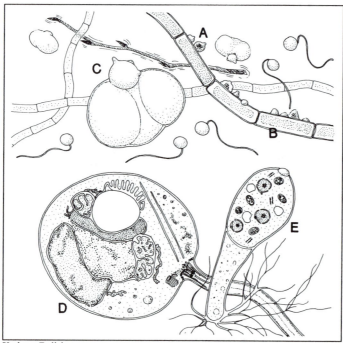

Kathryn Delisle

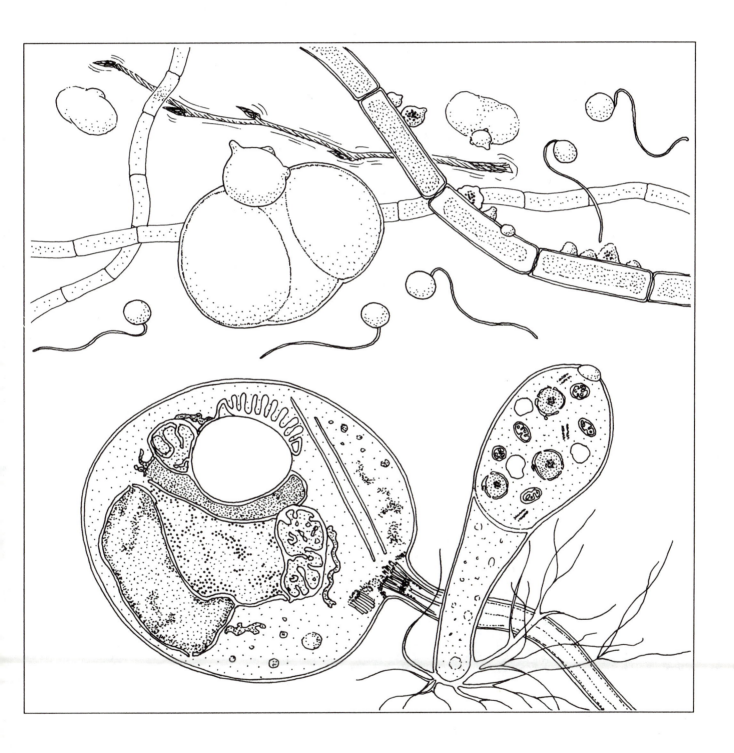

Pond Rocks

Phylum: Ciliophora. In brackish water, the open ocean, a small pond or puddle, the intestines of mammals, or even the sulfide-charged hot vents of the abyss, ciliates abound. Most ciliates are phagotrophic or osmotrophic microbes, but some, having acquired algal symbionts or borrowed chloroplasts, are secondarily photosynthetic.

Ciliates are a clearly distinguishable phylum of the Protoctista. They have rows of cilia called kineties, a cytosome (or cell mouth), and pronounced nuclear dualism (the diploid micronucleus is the germ nucleus whose meiotic products are exchanged during conjugation, whereas the larger macronucleus is the control center of the cell for protein synthesis and differentiation). The micronucleus does not divide by mitosis and has large quantities of uniquely organized DNA, as well as its own style of amitotic division.

Some 1,100 genera and 7,500 species of ciliates have been discovered. Most scientists estimate that at least 10,000 species are extant. Shown here beneath strands of the bladderwort *Utricularia* sp. (**A**)—which traps ciliates—are *Vorticella* (**B**) and *Stentor* (**C**) attached to rocks, free-swimming *Tetrahymena* (**D**), *Para-mecium bursaria* (**E**), and *Bursaria* sp. (**F**) containing prey *Paramecium*. The detail of *Tetrahymena* (**G**) shows the cytosome with specialized cilia, the locomotive cilia, the micronucleus and macronucleus, and the mitochondria.

Although some ciliates lack cilia (the shafts are so short that they are virtually nonexistent), none are without the underlying kinetosome. This detail of a ciliate kinetosome and cilium (**H**) shows the kinetosomal fibers and microtubular ribbons. Nearly all ciliates are single-celled, but some, such as the sessile *Acineta*, give rise to swimming forms, and others, like *Sorogena*, form spore stalks and are genuinely multicellular.

The huge (one meter wide) freshwater colonial ciliate *Ophrydium* looks like a mass of green algae, but it is a complex, light-loving ciliate formed by thousands of attached cells. If forced for more than two days to live in darkness, the colony throws off small green swimming cells that invade new well-lit habitats and begin to grow again into a colony. Tintinnids, a family of marine ciliates, form tests, hard parts that permit them to be recovered as fossils, even from Mesozoic sediments.

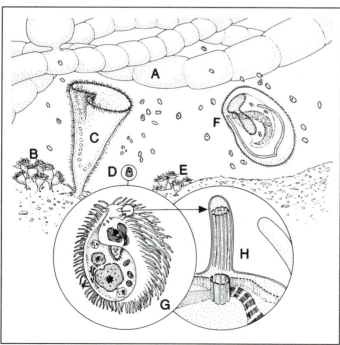

Kathryn Delisle

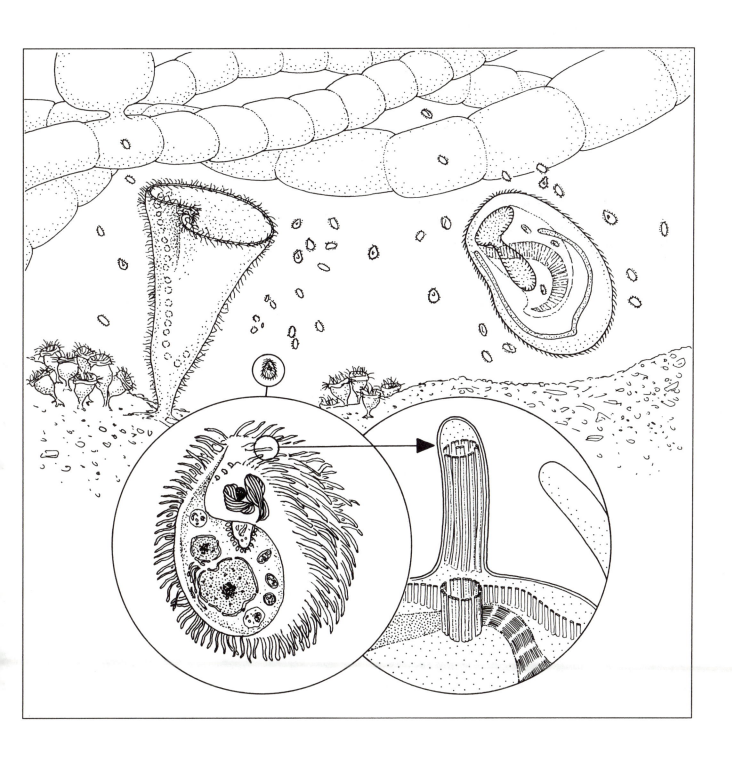

Tropical Coast

Phylum: Granuloreticulosa. Most granuloreticulosans are foraminiferans (called forams for short). Forams form a diverse group of marine organisms known for their elaborate tests (shells) of calcium carbonate or sand grains. Their fossil remains are useful to geologists, not only for oil exploration, but also in the reconstruction of ancient climates (paleoclimatology).

The largest forams are sometimes mistaken for snails, and even geologists who work with them refer to them as animals, as part of the fauna. They are clearly protoctists, entirely unlike animals, however, in their sexuality, reproduction, development, and cell structure. Either naked (e.g., reticulomyxids) or pore-bearing and shelled (foraminiferans), granuloreticulosans extend highly active leglike cell projections called reticulipodia. When these thin reticulipodia contact each other, they fuse to form the foram network.

Three representative genera of forams are shown here on a coral reef. *Heterotheca lobata* (**A**), with its anastomosing granular reticulopodia displaying two-way streaming motion, is shown in its diploid agamont form, which is capable of reproducing without sex (i.e., it is a stage in the life cycle that does not form gametes). The planktonic *Globigerinoides* sp. (**B**) has dinomastigote symbionts (see page 88) on its spines and rhizopodial network. The benthic foram *Textularia* sp. (**C**) has an agglutinated test made from surrounding sand grains.

These forams are pictured with their food organisms: green algae (see page 84), diatoms (see page 100), and ciliates (see page 94). Although most are predacious heterotrophs, many forams harbor symbiotrophic algae, or even foreign chloroplasts, and thus supplement their predatory food habits with photosynthesis.

Even though the forams are exceedingly diverse and show complex sex lives and developmental patterns (e.g., some burst to form hundreds of swimming cells at one time), the details of their biology are still poorly known. Attempts to grow forams in laboratory cultures tend to fail, in spite of their great abundance in marine and estuarine waters (in the open sea about forty living planktonic species are known). Forams are especially abundant in coastal sediments, where most of the thousands of living species reside. Some 4,000 living species have been identified; the fossil record boasts of 35,000 to 40,000 species.

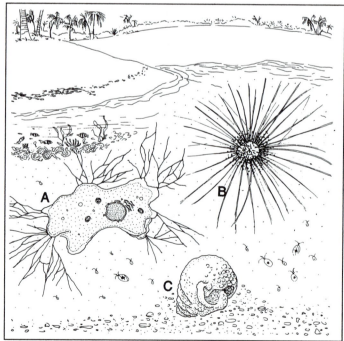

Kathryn Delisle

Mosquito and Blood

Phylum: Apicomplexa. Traditionally, people, domestic animals, or fish that were sick with or died of diseases that could be correlated with little dark bodies in the tissues and with cysts excreted in their feces were described as having sporozoan parasites. We now know that many unrelated protoctists form little dark spots in tissues as obligate symbiotrophs. They derive their nutrition from the blood of the chordate (see page 142) and in some cases are necrotrophic, causing serious disease. Most live nearly invisibly, however, having developed astounding ways to insure a stable food supply.

One great group of what were once called sporozoans contains the thousands of kinds of apicomplexans. Apicomplexans derive their name from a cell structure, the entirely modified anterior portion of the cell called the apical complex (**A**). These organisms live in the tissues of animals, notably arthropods (see page 144) and chordates (see page 142). The distinctive arrangement of fibrils, microtubules, vacuoles, and other organelles assures the forced entry of apicomplexans into the living bodies from which they derive their food.

Plasmodium sp., which is associated with malaria, is the most infamous apicomplexan. The life cycle of *Plasmodium* involves its forced entry into two animal species types: humans and the *Anopheles* mosquito (**B**). The life-cycle stages of *Plasmodium* shown here in the intestinal lining of the mosquito (**C**) are the undulipodiated zygote (**D**), the unsporulated oocyst (**E**), and sporozoites excysting from the oocyst (**F**).

Once excysted, the sporozoites enter the mosquito's saliva and pass through its proboscis (**G**) into the human bloodstream. In the human vein, the sporozoite attaches to a red blood cell (**H**), which grows into the vegetative stage called the trophozoite, which then breaks up, releasing merozoites (**J**), which are taken up with the blood by a mosquito. Not shown here is meiosis, the process of cell division that reduces the number of chromosomes from the diploid (double set) to the haploid (single set) form. Meiosis in *Plasmodium* occurs in the mosquito.

Other well-known apicomplexans include *Toxoplasma,* which infects people via their cats, and *Eimeria* (as well as other coccidians), which lives in farm animals and pigeons.

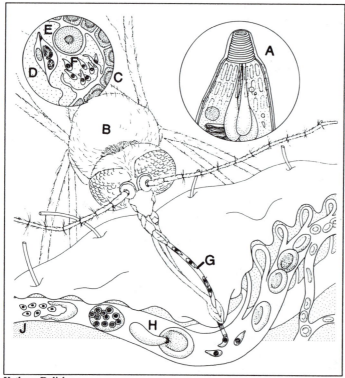

Kathryn Delisle

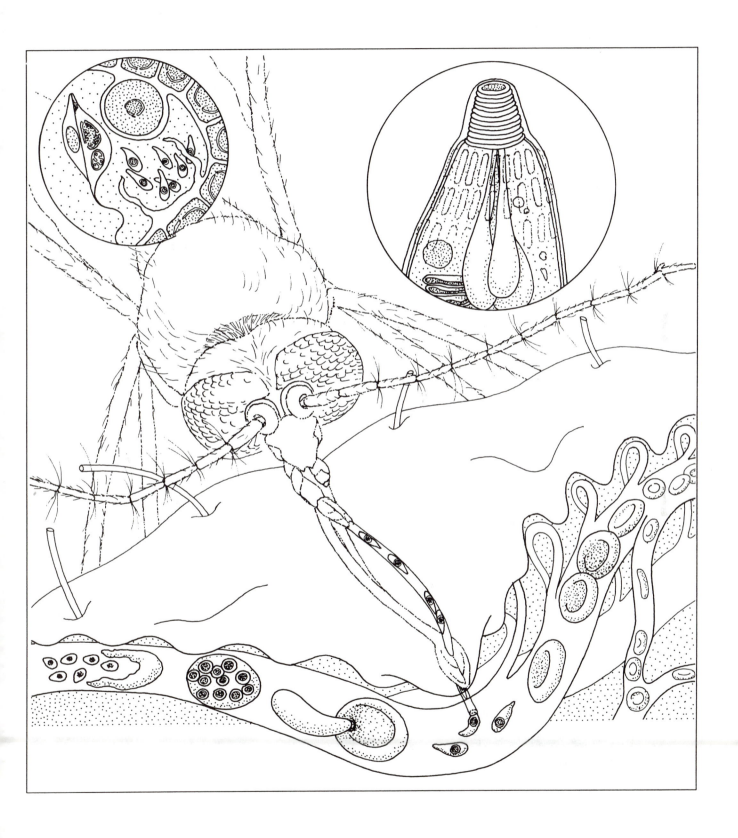

Oceanside

Phylum: Bacillariophyta. Known as diatoms, these organisms are sexual protoctists. Primarily single-celled, aquatic, photosynthetic organisms, all of them form elaborate and beautiful tests (shells) of silica. Some form complex colonies. Diatoms live most of their life cycle in the diploid state; then they produce haploid gametes. In one group (the centric diatoms), motile male sperm is undulipodiated and fertilizes an immotile, naked, female ameboid cell called a protoplast. In the other group (the pennate diatoms) both sex cells are ameboid; neither has undulipodia. They shed their silica walls as they emerge and fuse.

Diatoms have greenish brown chrysoplasts and are masters of biomineralization, depositing silica in gorgeous patterns in their cell walls. Their empty tests, mostly as fossils, are called diatomaceous earth. Diatoms are used extensively in water filtration. They also display a distinctive form of actin/protein-based locomotion from secretions through slits in their cell walls.

Pictured in this marine habitat is the green alga (a chlorophyte) *Cladophora* (**A**), with attached diatoms (*Rhoicosphenia, Gomphonema,* and *Cocconeis,* **B**); the diatoms *Opephora, Amphora,* and *Fragilaria,* attached to a sand grain (**C**); and the centric diatom *Melosira* sp., with a cross section showing its plastids, nucleus, and mitochondria (**D**), its sperm (**E**), and a detail of the spines lining its two valves (**F**).

Some diatoms are colonial, living stacked on top of each other. Others are internal symbionts of foraminiferans (see page 96). About 250 genera and some 100,000 species of diatoms have been described and named, of which 10,000 are living. Diatoms are said to be the second most abundant type of organism worldwide. Bacteria, of course, are the most numerous.

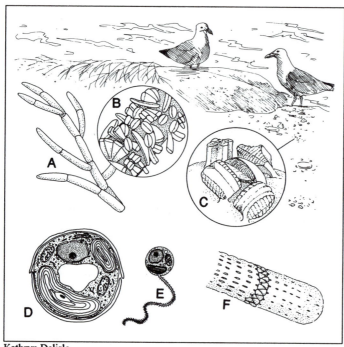

Kathryn Delisle

Pond Bottom

Phylum: Oomycota. The oomycotes, another fungus-like group of protoctists, are often referred to as water molds. A better name might be "egg protoctists" or "egg molds." They differ from chytridiomycotes (see page 92) and hyphochytrids (see page 82) in many ways. Traditionally, oomycotes have been classified as fungi because in damp places they form visible, fuzzy, white masses. Their cell walls are made of cellulose, and their zoospores have two undulipodia. One undulipodium is directed forward during swimming and has mastigonemes; the other is smooth and trails behind.

Sexuality in oomycotes is unique, complicated, and distinctive. *Saprolegnia parasitica* (**A**)—which lives osmotrophically on the yellow perch, *Perca flavescens* (**B**), and its eggs—displays a typical oomycote sexual cycle (**C**). The ends of the filaments, specialized hyphal tips, develop into female oogonia and male antheridia (**D**). The antheridia and oogonia fuse; then the male nuclei can migrate. The male nuclei enter and fertilize the oosphere, or egg (**E**). This large cell resembles an animal egg. The fusion results in the formation of oospores (**F**), which divide mitotically. Their products are diploid zoospores (**G**) that swim away and propagate the water mold. Oospores also undergo meiosis and form haploid nuclei that then differentiate into haploid zoospores, which are released. Alternatively, zoospores are produced by direct mitotic division in a sporangium (**H**).

Oomycotes usually live necrotrophically on plants or freely in fresh water on plant debris. The group includes such infamous plant pathogens as *Phytophthora infestans,* associated with late blight of potato, and *Plasmopara viticola,* associated with grape mildew.

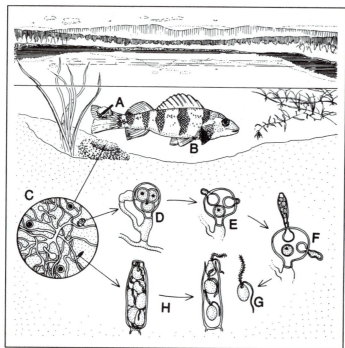

Kathryn Delisle

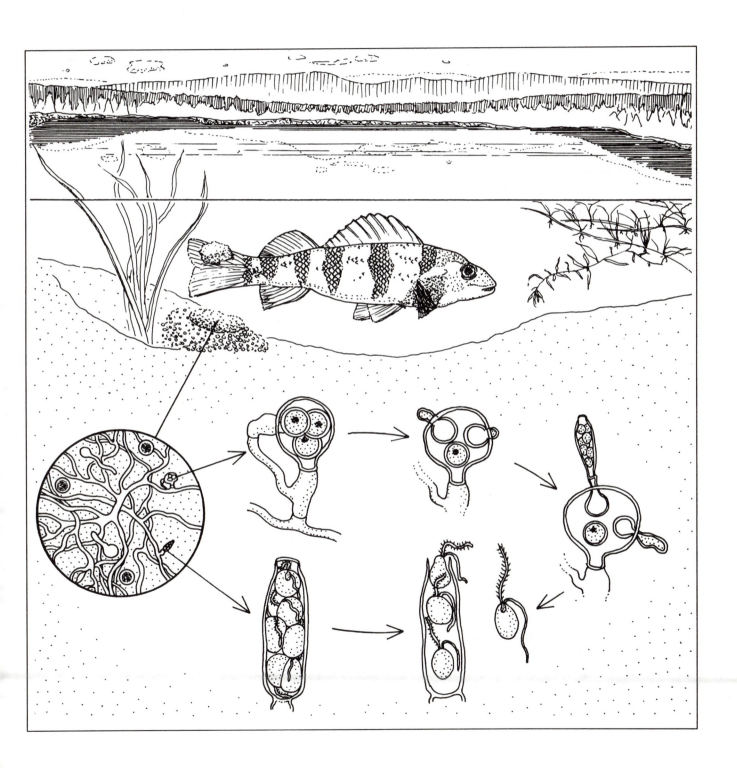

Pacific Nearshore Waters

Phylum: Ellobiopsida. The ellobiopsids are obscure but important symbiotrophs of planktonic animals—shrimp, copepods, amphipods, and euphasids (all arthropods, see page 144). Although their life history is not fully known, observed behavior of one of the well-known species, *Thalassomyces marsupii*, a symbiotroph with the amphipod *Parathemisto* (**A**), is shown here.

The ellobiopsid attaches to the reproductive organs—the ovaries or egg masses—of the amphipod and grows out of the animal's body, forming numerous branches, or trophomeres (**B**). These branches then develop into coenocytic (multinucleate), reproductive structures called gonomeres (**C**). This detail shows the outside of the gonomere and a cross section with the many nuclei inside. The gonomere facets, or breaks into many faces, which open and release unicellular spores (**D**). These spores develop two undulipodia—one posterior and one circumferential (**E**). It is suspected that these motile "zoospores" then enter another animal to establish the symbiosis again. This second invasion by un-dulipodial zoospores has not yet been observed or confirmed experimentally. Further research is needed to determine to which protoctist phylum these organisms are most closely related. Until that time this group is considered *incertae sedis* (literally, "without a seat").

Phylum: Ebridians. Because of their skeletons, ebridians are known better to marine paleontologists than to biologists. The ebridians are a small group of zoomastigotes with internal skeletons composed of siliceous rods. They live in coastal marine environments and form periodic "blooms" following the blooms of their food organisms: diatoms (see page 100) and dinomastigotes (see page 88). *Ebria*, a common genus, is shown here. *Ebria tripartita* (**F**) displays the two distinctive anterior undulipodia of the phylum. Also shown is the siliceous skeleton (**G**) and cell division (**H**). The name *Ebria* comes from the Latin word *ebrius*, which means "drunken"—a reference to the appearance of this organism when it is swimming.

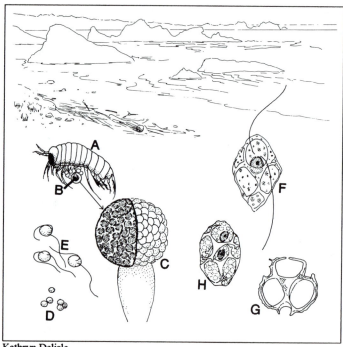

Kathryn Delisle

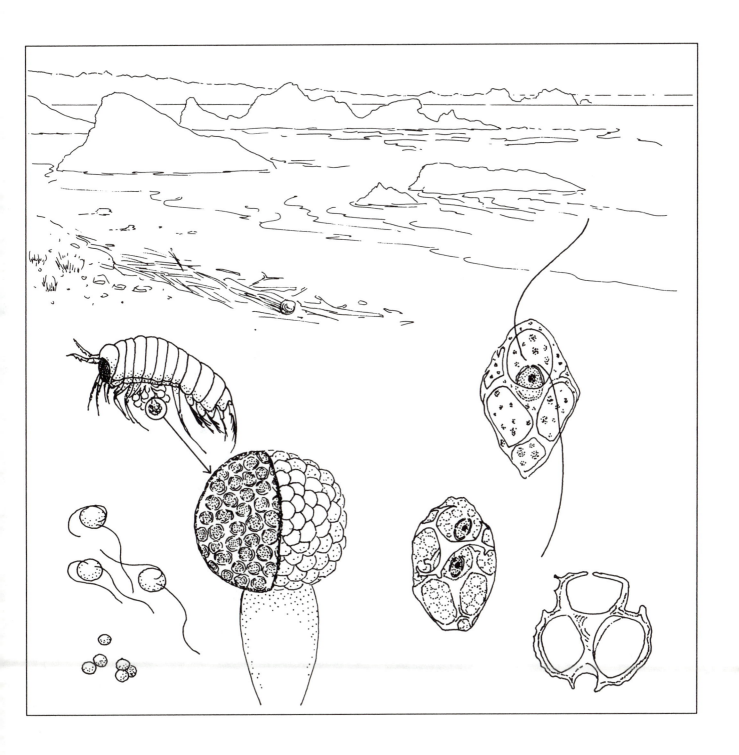

Chapter 3: Fungi

The kingdom Fungi includes molds, mushrooms, yeasts, and other eukaryotes that form fungal spores (Figure 12). Fungi lack undulipodia at all stages of their lives. If fungi indulge in sex, they conjugate: Their threadlike or spherical bodies fuse to form recombinant haploid (or dikaryotic) new bodies. These fascinating eukaryotic organisms can produce huge numbers of hardy propagules called spores without engaging in sex at all.

Fungal spores germinate to grow into slender tubes called hyphae. Crosswalls called septa divide hyphae into cell-like units in some fungi. In other fungal species, walls are not formed. Each cell-like unit usually contains more than one nucleus and many mitochondria (the number depends on the species of fungus). Septa seldom completely separate the cells; thus, cytoplasm flows more or less freely through the hyphae.

The growing form of most fungi is a large mass of hyphae called a mycelium. Reproductive structures, which are also made of hyphae, form in response to seasonal changes in weather: rain, desiccation, etc. Some fungal reproductive structures include puffballs, morels, mushrooms, and shelf fungi. Among the largest and most complex fungal reproductive structures are the immense shelf fungi on trees, some of which arise from mycelia that are up to two meters in diameter. Most fungi, however, are microscopic.

All fungi are heterotrophs that absorb food through their chitinous cell walls, rather than ingesting (eating) it. Nearly all are aerobes. Fungi secrete powerful enzymes that break down live or dead materials into a solution of molecules outside the fungus body. These sugar, protein, and other molecules are transported into the fungus through its cell walls. The stiff cell walls of fungi and the exoskeleton (shell) of lobsters both contain chitin, which is a tough, nitrogenous, long-chain polysaccharide. Fungi are tenacious, resisting cold and hot weather, severe desiccation, and other environmental insults. Their cell walls resist water loss. Some even grow in strong acid. Others survive in clean tap water and other environments that contain nearly no nitrogen. Still other fungi are capable of etching glass lenses as they grow on minute quantitites of organic compounds. Others contaminate jam and even strong chemicals with their fuzzy growth. Often the least conspicuous of the eukaryotes, fungi are among the most resilient, living in every ecosystem from Antarctic rocks to human skin.

Truffles and chanterelles are prized edible fungi. Fungi are employed to ripen Camembert, Roquefort, and other blue cheeses. Fungi are treasured as the original source of the antibiotic penicillin. Fungi are also the source of the medicine ergot, which constricts smooth muscle. In small doses ergot derivatives constrict arteries and are given to relieve migraines and after childbirth to prevent hemorrhaging. In large doses, however, ergot is toxic.

The human body harbors mold- and yeastlike fungi; some normally cause no problems. Suppression of the immune system by certain medications, organ and tissue transplant procedures, radiation, malnutrition, debilitating illness, AIDS, or leukemia can allow fungi to invade human tissue. Artificial body parts from hips to heart valves provide growth sites for fungal colonies, thereby increasing the risk of fungal infections. Fungi harbored within "sterile" urinary catheters are a common source of nosocomial (hospital-acquired) infection. Thus, the prevention and treatment of fungal infestations in humans is growing in importance.

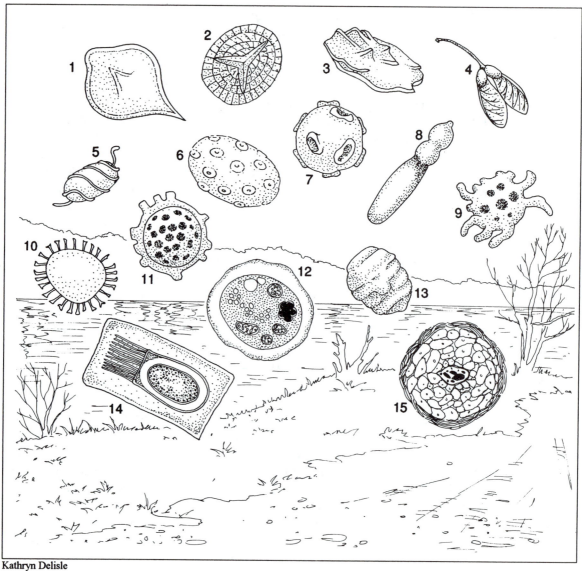

Kathryn Delisle

Figure 12

Pictured here are several types of spores and seeds, both of which are resistant propagules: (1) ascospore (mold), (2) fern spore (nonflowering plant), (3) deuteromycote spore (fungus), (4) red maple samara, *Acer rubrum* (seed of flowering plant lies within), (5) horsetail spore, *Equisetum arvense* (nonflowering plant), (6) ascospore (fungus), (7) moss, *Polytrichum* sp., (8) basidiospore, (9) lichen soredia (propagule consisting of algal cells surrounded by fungal hyphae), (10) dinomastigote cyst (protoctist), (11) basidiospore (fungus), (12) ameba cyst, *Paratetramitus* (protoctist), (13) tardigrade tun (animal, see page 116), (14) *Arthromitus* parent cell with walled spore with attached spore filaments (bacterium), and (15) macrocyst (giant cell of cellular dictyostelid slime mold, protoctist).

Spores are small, usually microscopic, propagules containing at least one genome. At some time, spores can mature into the growing form of the organism. Spores often resist starvation or heat. "Spore" is a general term that has been applied to spherical propagules from the moneran, protoctist, fungal, and plant kingdoms. A spore from an ameba of a cellular slime mold and a cyst from the same species may appear identical in external structure. Two spores, on the other hand—one from an aplicomplexan and the other from a basidiomycote, for example—will differ entirely from each other in genetic information.

Seeds are fertilized ovules of seed plants (Gingkophyta, Coniferophyta, Gnetophyta, and flowering plants, such as the maple shown here, 4) that usually include a food supply such as endosperm. Seeds propagate seed plants.

Orchard

Phylum: Zygomycota. The zygomycotes, or mating molds, are mostly filamentous organisms that live off of substances rich in organic compounds dissolved in the water or tissues of animals, plants, or protoctists. The fungal filaments (hyphae) clumped together as a visible mass, a fungal body, are called the mycelium. Some zygomycotes are predatory: Shown here are *Dactylaria* sp. (**A**) attacking nematodes (see page 134) and *Dactylella tylopaga* (**B**) attacking amebas (see page 52) in the soil.

After the sexual fusion of hyphae that are of complementary genera and are thus able to mate, the fused hyphae form resting structures called zygosporangia. They result from the fusion of two swollen structures and are multinucleate. Mature zygosporangia are visible to the naked eye as tiny dark dots. Zygomycotes also produce sporangia without any mating or fusing, which are raised on stalks called sporangiophores. These sporangia are visible as black fuzz on bread (e.g., black bread mold, *Rhizopus*).

A small group of cosmopolitan or widely distributed zygomycote genera are crucial to establishing healthy environments for plant roots: The fungi provide the plants with phosphorus from rocks and soil. Unable to photosynthesize, the fungi feed osmotrophically on carbohydrates derived from the photosynthesis of the plants in which they reside. The zygomycotes are the fungal symbionts of the endomycorrhizae and reside within plant roots. Fungi such as *Glomus* and *Endosporella* live symbiotically with 80% of all vascular plants.

Some have thousands of nuclei, which in synchrony fuse in pairs in what looks like an incredible orgy. Mucormycosis occurs when the zygomycote *Mucor* infects people with leukemia or AIDS.

Phylum: Ascomycota. The most familiar molds are ascomycotes, and the parts we usually notice are the clumps of sexual reproductive organs, such as the delicious morel, *Morchella esculenta* (**C**). The ascomycotes get their name from the ascus, a saclike structure containing haploid ascospores, which are not found alone but are often bundled together. Ascospores are found in all ascomycotes, whether they are the filamentous forms or the subvisible single-celled yeasts.

The morels, which are among the choicest edible fungi, are in the class Euascomycetae, which also includes the fungal partners of lichen symbioses. The yeasts ferment sugar or starch as nutrient and are used in the production of bread, beer, wine, and a yogurtlike beverage popular in eastern Europe called kefir.

Ascomycotes produce propagules in prodigious numbers; many survive cold and dryness very well. These propagules form asexually (they have only one parent) and are called conidia. Primarily conidia are responsible for the propagation of these widespread fungi. Unlike bacterial spores, fungal spores cannot survive boiling, but some can be dry for years and will grow vigorously if organic-rich water comes their way. More than 30,000 species of ascomycotes have been described, including the necrotrophs like the fungi associated with certain diseases of trees: chestnut blight (*Endothia parasitica*) and Dutch elm disease (*Ceratocystis ulmi*).

Phylum: Mycophycophyta. Members of this fungal phylum are called lichens, which are symbiotic associations of ascomycotes (see this page) and green algae (see page 84) or cyanobacteria (see page 34). There are about 20,000 species of lichen fungi, almost all of which are ascomycotes that differ in appearance from the 30,000 ascomycotes described above. The lichen shown here is *Parmelia conspersa* (**D**). In cross section (**E**), the algal and fungal partners are visible. The photosynthetic algal component provides the fungus with nutrients; the fungus retains water from the symbiotic complex and modulates the sensitivity of the partnership to environmental change. Lichens reproduce by soredia (see Figure 12, page 108). Lichens also disseminate via bits blown by wind to new locations.

Kathryn Delisle

Forest Clearing

Phylum: Basidiomycota. Included in the basidiomycotes are many of the most familiar larger fungi, including the mushrooms, puffballs, coral fungi, and the rust and smut plant pathogens. All make basidia, microscopic clublike structures tucked away in gills or pores. The basidia release haploid basidiospores, which are the end product of a complex cycle of fused hyphae. In the heterokaryotic mycelium (one with several different genetically distinguishable nuclei in the same hypha), the nuclei finally fuse (mate), and immediately after, they divide to form spores for the next generation.

The two classes of basidiomycotes are the Heterobasidiomycetae (the necrotrophic rusts and smuts) and the Homobasidiomycetae (which include most of the mushrooms). Pictured here are *Amanita* (**A**), a gilled mushroom, some species of which are among the most poisonous mushrooms; *Fomes* (**B**), whose mycelium extends around the roots and covers the small rootlets of trees with a fuzzy sheath; *Armillaria* (**C**), the "honey mushroom"; and *Boletus* (**D**), whose edible mushroom contains pores rather than gills (**E**).

Thousands of basidiomycote species form ectomycorrhizae, symbioses that provide phosphorus to the plants and carbohydrates to the fungi. The nearly ubiquitous zygomycote-based endomycorrhizae (see page 110) grow well on most plants but especially well on broadleaf trees and herbs. Ectomycorrhizae, by contrast, are found mainly on trees and shrubs.

Phylum: Deuteromycota. The deuteromycotes are molds that produce single-parent spores (conidia), usually in great profusion. No sexual act has ever been witnessed in the deuteromycotes, but they are thought to be ascomycotes (see page 110) or basidiomycotes (see this page) that have lost their potential to differentiate asci or basidia. Traditionally, deuteromycotes have been called the Fungi Imperfecti because they do not form sexual structures—i.e., botanically speaking they are not perfect. Some deuteromycotes have exhibited a parasexual cycle in which unspecialized hyphae fuse and their nuclei recombine.

Pictured here (**F**) is *Titaeosporina* forming an acervulus on a leaf surface. Other deuteromycotes include the famous antibiotic producer *Penicillium; Aspergillus,* the fermenter used in making soy sauce and also the fungus responsible for aspergillosis, a form of pneumonia; *Candida albicans,* associated with a common vaginal infection and found in immunosuppressed adults; and *Trichophyton,* which lives in skin infected with athlete's foot and groins afflicted with jockstrap itch. Cryptococcosis and histoplasmosis are other infections caused in immunosuppressed humans by deuteromycotes. *Rhizoctonia,* a bane of gardeners and farmers, is a soil-dwelling deuteromycote associated with root rot of plants.

Donna Reppard

Chapter 4: Animalia

Animals run the gamut of biodiversity from velvet worms (see page 144), which prowl the rain forest floor, to the longfin char, which inhabits one of the iciest, most remote continental lakes on Earth. The tiniest animals are microscopic. Among the smallest are aquatic rotifers (see page 134), 0.04 millimeters long. The immense whales, members of our own phylum—Chordata (see page 142)—are the largest present-day animals.

Most animal species are aquatic; of these, most dwell in shallow waters. Cold, deep waters are some of the most recently explored ecosystems; thus, their animals are some of the most recently described. The longfin char (*Salvethymus svetovidovi*) was discovered in 1985 in a three-million-year-old freshwater ecosystem housed in a meteorite impact crater in Siberia. Tube worms taken from a depth of 2,500 meters off the Galapagos Islands belong to one of the most recently described animal phyla—Vestimentifera (see page 146).

Soil-inhabiting animals such as annelids (see page 136) and nematodes (see page 134) depend on aqueous environments for their entire life cycle. Few animal species are genuine land-dwellers: arthropods (see page 144)—crickets, spiders, and butterflies—and chordates (see page 142)—birds, reptiles, and mammals.

With few exceptions, animals take food into their bodies (i.e., they have ingestive nutrition). Animals have elaborate schemes for ingesting food. The longfin char ingests other members of the ecological community of Lake Ll'gygytgun ("ice lake"), such as zooplankton and amphipods. Sponges engulf food particles in phagocytotic cells. Leeches suck blood from unfortunate swimmers who venture into warm ponds. Arboreal mice of tropical forests lap flower nectar. Some aquatic animals absorb food molecules across their cell membranes. Humans consume an enormous variety of foods, such as pizza, grapefruit juice, and sweet corn.

In our classification system, then, animals are eukaryotic, heterotrophic, multicellular, diploid organisms that develop from two different haploid gametes, sperm and egg. These two haploid cells join in the process of fertilization and produce a diploid zygote that divides to form a blastula embryo. The blastula is a liquid-filled sphere of cells that is characteristic of members of the kingdom Animalia only. Common patterns of development reflect what is probably the common ancestry of members of different phyla. An example is the larval swimming stage of molluscs (see page 142), sipunculans (see page 140), and annelids (see page 136). Larvae (Figure 13) aid in the spread of young animals and then metamorphose into a diversity of adult forms. By definition, adults are

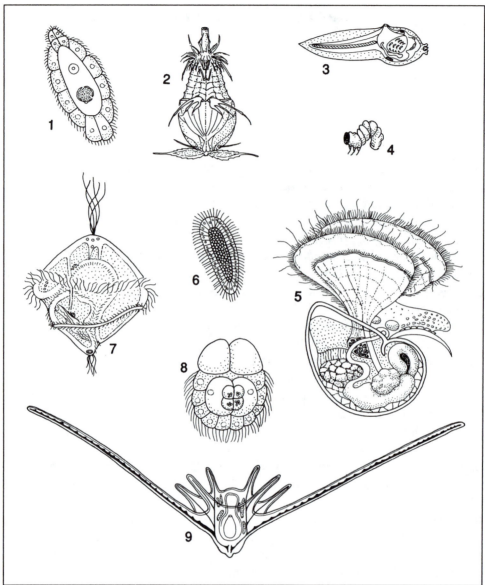

Kathryn Delisle

Figure 13

A larva is an immature or resistant stage in the postembryonic development of an animal. Larvae differ considerably from their corresponding adult stages in form, function, and habitat. Most larvae are aquatic; often they are swimmers that disperse the young animals. The similarities of larvae of many classes of animals may well have more to do with the power of dispersal of these developmental stages—trochophore, pluteus, or veliger—than with their ancestry. Some are resistant to environmental insults such as desiccation, heat, and cold. Several types of larva are shown here with their associated phyla: (1) vermiform (Mesozoa), (2) Higgin's larva (Loricifera), (3) tunicate tadpole (Chordata), (4) beetle grub (Arthropoda), (5) veliger (Mollusca), (6) planula (Cnidaria), (7) trochophore (Annelida), (8) infusoriform (Mesozoa), and (9) pluteus (Echinodermata). The tun of the water bear (or tardigrade) is not a young larva; rather it is an extremely resistant stage that develops in response to low environmental temperature (see Figure 12, page 108).

the forms that produce sex cells—sperm and eggs—that fuse to start the life cycle of the animal again.

Like all other animal cells, those of the blastula lack the cellulose-containing cell walls that are characteristic of plant cells. Moreover, animal cells have complex cell connections. Elaborate gap junctions and desmosomes ensure and regulate the flow of materials, as well as communications and support, between cells. Multicellularity does exist in all five kingdoms. Animal cells are further linked into tissues—char muscle, pentastome epithelium, echiuran nerve cord. Only placozoans (see page 118) and sponges (see page 120) lack tissues.

Animals traditionally have been subdivided into two large groups: "them"—those lacking backbones, or "invertebrates"—and "us"—those with backbones, the "vertebrates." Vertebrates include all members of the chordates (birds, amphibians, mammals, see page 142) except urochordates (e.g., tunicates like *Lissoclinum,* see page 36) and cephalochordates, a strange group that contains fishlike forms of marine life. Nearly all members of the first group (thirty-three phyla and many members of our own phylum Chordata) are marine animals. Most of these phyla have no land-dwelling species at all. The vast silent majority (since nearly all members of the kingdom are dismissed as "invertebrates") tend to be subdivided further into radially symmetrical animals, such as jellyfish and comb jellies, and the rest, which are bilaterally symmetrical.

Contemporary classification schemes of animals are based on evidence derived from fossils and from extant organisms: molecular biology, chromosomal cytology, pattern of embryonic development, behavior, and morphology. The most likely ancestors common to living animals probably evolved during Ediacaran times over 600 million years ago (see Figure 11, page 50). Most biologists agree that animals evolved from the choanomastigotes (see page 72), microscopic protoctists ancestral at least to the sponges. This protoctist/animal connection is deduced from the fine structure of sponge and choanomastigote cells.

In what environment did animals evolve? This question continues to be debated. The relationships between animals, plants, fungi, and other eukaryotes inferred from the comparison of ribosomal RNA sequences shows that the animal lineage includes choanomastigotes. This evidence also confirms the divergence of sponges (see page 120) and comb jellies (see page 124) from other animal phyla and implies that fungi and animals share a more recent ancestor in common than either kingdom shares with plants. The origin and diversity of animals and their relation to other kingdoms will be explored vigorously as new phylogenies emerge.

Pebbled Sea Bottom

Phylum: Placozoa. Probably the simplest animals alive today, placozoans change shape as they creep. They look like large, lumpy amebas (**A**). Only one placozoan species, *Trichoplax,* is known. (By contrast, the phylum Arthropoda encompasses at least a million species). Lacking distinct organs (such as kidneys or hearts) and tissues (such as nerve and muscle), *Trichoplax* (along with sponges, see page 120) is not considered part of the main line of animal evolution.

Why then is this tiny, multicellular, marine organism with neither head nor tail, neither left side nor right side, classified as an animal? The embryo of *Trichoplax* is the hollow ball of cells (a blastula) that is characteristic of all animals and that results from eggs fertilized by sperm. In addition to sex, *Trichoplax* can reproduce by splitting into two amebalike individuals, each containing about a thousand cells. Like many sea animals, very small placozoans can swim and so probably disperse the species. Large adults crawl using their ventral cilia.

Phylum: Kinorhyncha. Kinorhynchs (from the Greek words *kinein,* meaning "to move," and *rhynchos,* meaning "snout"), such as *Echinoderes kozloffi* (**B**), are obscure marine worms with spiny heads. A kinorhynch moves forward by forcing fluid into and thereby extending its head, then anchoring the spines along its body, hauling itself forward, and finally retracting its head. Most kinorhynchs are free-living but nonswimming

inhabitants of muddy ocean floors and coarse sediments as far north as Greenland. Some may be commensal, living in sponges (see page 120) and attached to members of other marine phyla. Kinorhynchs usually are brownish yellow. The segmented cuticle differentiated into plates and recurved spines that is common to kinorhynchs and loriciferans (see this page) indicates that these two phyla most likely have descended from a common ancestor, even though no fossils of either phylum are known.

Individual kinorhynchs are male or female, but usually no external features distinguish the sexes. A spermatophore (i.e., a special packet of sperm) is deposited in females of some species to fertilize the eggs. The fertilized eggs develop externally. As young kinorhynchs develop into adults, the juveniles molt their cuticles at least six times to accommodate their growing bodies.

Phylum: Loricifera. Like all other animals, the loriciferan shown here, *Pliciloricus enigmaticus* (**C**), develops from a blastula (**D**). *P. enigmaticus* can telescope its mouth cone, head, neck, and thorax into the girdle of six sculptured spiny plates called the lorica. All loriciferans bear a lorica and anchor themselves to submarine rocks and to other animals among the sand grains. Loriciferan larvae, such as the *Nanaloricus mysticus* larva shown here (**E**), swim by using their leafy toes as paddles. Adults can move across subtidal sediment particles via their ventral spines.

The natural history of loriciferans is known only poorly. For example, we do not know what they eat or how they feed, and we have not yet unraveled their sex lives. The spiny cuticles and retractable mouth cone of loriciferans and kinorhynchs (see this page) probably are homologous, meaning that they evolved from a common ancestor.

Christie Lyons

Open Ocean Rock and Sand Bottom

Phylum: Porifera. The walls of this Venus flower basket sponge, *Euplectella speciosissima* (**A**), are penetrated by thousands of pores. Although sponges lack a digestive tract, nerves, and respiratory and circulatory systems, they have evolved an elegant mode of feeding. Plankton (free-floating aquatic organisms) and detritus raining down through the sea are drawn into the pores of the sponge, then wafted along inhalant canals by microscopic beating organs of motility called undulipodia, borne by collar cells. Once inside, food particles are engulfed by ameboid cells lining the canals.

Symbiotic algae provide nutrients and oxygen to many species of sponges, as well as absorbing their wastes. The algal symbionts move from mature sponge to offspring by hitching a ride on the gemmules (particles made up of cells) by which sponges disperse. Sponges also reproduce with sperm and eggs. Sponges are colored by algal symbionts or by their own pigments and may be any color in the rainbow—red, orange, yellow, green, blue, purple, brown, or white. Some sponges are bioluminescent: They glow in the dark.

Phylum: Chaetognatha. Flips of the tail send chaetognaths (from the Greek for "hairy jaws"), or arrow worms, darting through the open ocean on migrations to the depths during the day and to the upper ocean at night. Fins stiffened by rays stabilize them. Deep-water chaetognaths are tinted pink, red, or orange. Fossils indicate

that arrow worms, such as *Sagitta bipunctata* (**B**), had evolved by the Carboniferous period, when large woody club mosses and treelike horsetail plants dominated great forests. The developmental pattern of arrow worm embryos shows that they are related to chordates (see page 142), echinoderms (see page 128), hemichordates (see this page), and vestimentiferans (see page 146). In animals of all these phyla, the mouth is at the opposite end of the embryo from the anus, and the opening into the hollow ball of cells (blastula) that is the early embryo eventually becomes the anus of the adult animal.

Phylum: Hemichordata. One type of hemichordate, the acorn worm (such as *Ptychodera flava,* **C**) burrows into the sea bed with its acorn-shaped, muscular proboscis. Some acorn worms line the burrow with mucus, to which food and sediment stick; cilia that cover the soft body of the acorn worm transport the food mass to its mouth, at the base of the proboscis. Other hemichordates ingest sediments, from which they extract nutrients. The type of larva that lacks strong swimming ability, called tornaria, is characteristic of many hemichordate species. Other hemichordates, such as *Saccoglossus kowalevskii*, develop directly from egg to young worm, bypassing the larval stage. Tornaria are free-living plankton, those small animals and algae that are dispersed by water movement.

Hemichordate species that secrete rigid tubes and possess tentacles are called pterobranchs (from the Latin for "feather gills"). Acorn worms (class Enteropneusta) may have descended from pterobranchs. What evidence links hemichordates to chordates (see page 142), including humans? Embryos of hemichordates and chordates have a common pattern of development: The blastopore opening of the embryo develops into the adult anus. Most hemichordates and chordates are also characterized by gill slits. In humans, the gill openings eventually become, among other things, tiny middle-ear bones.

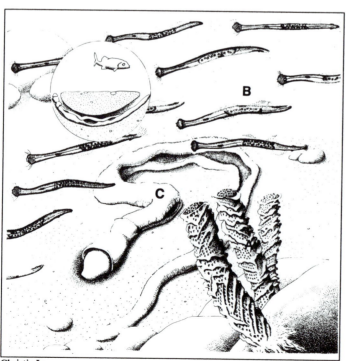

Christie Lyons

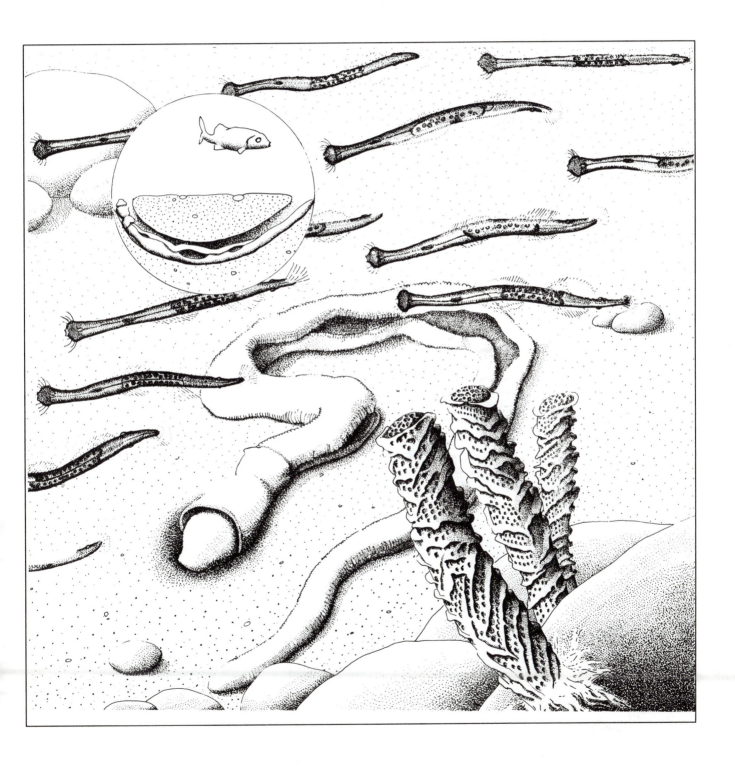

Caribbean Reef Seafloor

Phylum: Cnidaria. Known also as coelenterates, cnidarians (from the Greek word *knide,* meaning "nettle") include the well-known coral and jellyfish. The four major classes of cnidarians are the Anthozoa (which includes sea anemones and corals), the Hydrozoa (hydras), the Cubozoa (sea wasps and other cuboidal medusas), and the Scyphozoa (marine jellyfish such as *Obelia,* illustrated here).

There are two forms of cnidarians: colonial polyps (**A**) and medusas (**B**). The polyps produce reproductive bodies (**C**) at their branches and have a larval form (**D**). An immature colony (**E**) is shown growing on the sandy ocean floor. In hydrozoans, which are more likely than anthozoans or scyphozoans to be found in fresh water, the hydralike polyps predominate the medusa forms. Anthozoans have a flowerlike polyp but no medusa.

All cnidarians have special stinging lasso cells on their tentacles and body wall. Tentacles dangling from the four corners of the cuboidal medusa of the sea wasp can inflict serious damage on humans. After shocking and immobilizing small prey, the tentacles pass the dead victim, now food, to a hollow digestive cavity called the coelenteron, which is the defining feature of all members of the phylum.

Coral reefs are generation upon generation of cnidarian skeletons, cemented by the limestone secretions of cnidarian polyps, that act as fortresses.

Christie Lyons

Open Ocean

Phylum: Ctenophora. Multicolored lights flashing in the night sea often are the visible signs of bioluminescence of free-swimming ctenophores (from the Greek words *kters,* meaning "comb," and *pherein,* meaning "to bear"), or comb jellies. The beating of their comb plates (bands of fused cilia that run lengthwise along their bodies) propels comb jellies through the ocean. Eight rows of combs distinguish comb bearers from other jellylike, radial animals with a digestive cavity with only one opening (a coelenteron), such as jellyfishes and hydras (see page 122). These ocean floaters capture live prey with a pair of trailing, sinuous tentacles, then retract their tentacles and wipe food off on mobile lips. Adhesive lasso cells cover their tentacles; the touch of prey triggers the sticky cells.

After a storm glossy cat's eyes, sea gooseberries, sea walnuts, and Venus's girdles—all comb jellies—appear on the beach. Divers in submersibles collect comb jellies with slurp guns, observing that ctenophores are the most abundant planktonic animals at 400 to 700 meters below the surface. Several genera of ctenophores are shown here: the lobate comb jelly *Bolinopsis infundibulum* (**A**); Venus's girdle, *Cestum veneris* (**B**); the sea gooseberry, *Pleurobrachia pileus* (**C**); and the sea cucumber, *Beroe cucumi* (**D**).

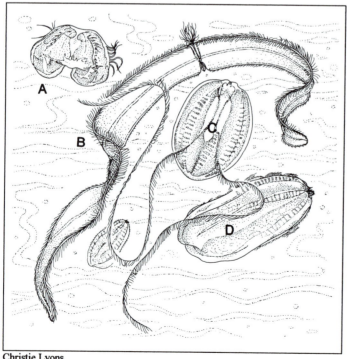

Christie Lyons

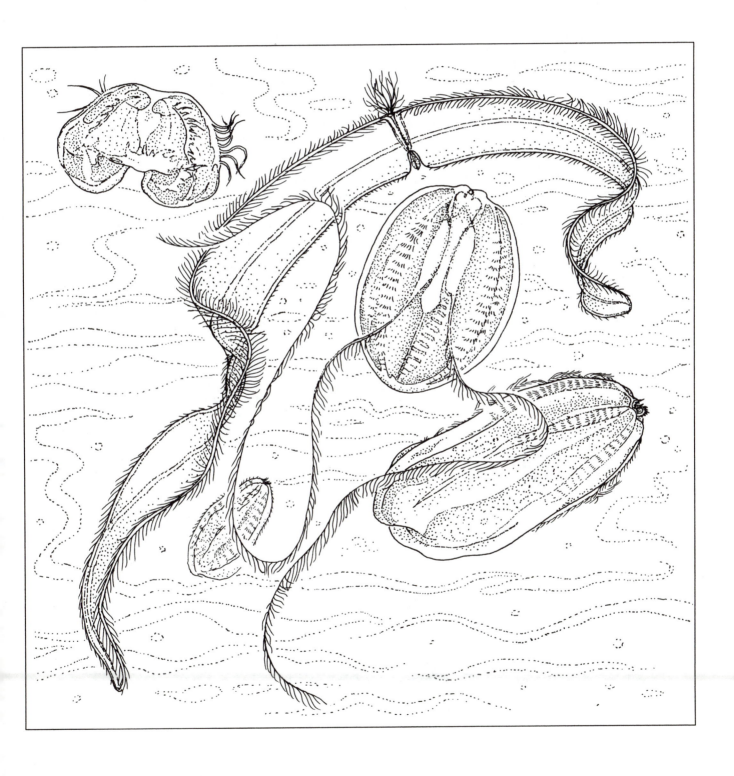

Philippine Coral Sands

Phylum: Mesozoa. A mesozoan (from the Greek *mesos,* meaning "middle," and *zoion,* meaning "animal") has only one organ, a gonad. Because mesozoans absorb nourishment from the urine of the animals that they inhabit (such as *Dicyema truncatum,* **A,** which lives symbiotrophically in *Nautilus,* **B**), these symbiotrophs function adequately without their own circulatory, skeletal, muscular, excretory, digestive, and nervous systems. In shallow coral sand, such as this community near the Philippine Islands, mesozoans are widespread in squid, octopods, and other cephalopod molluscs (see page 142).

Sexual adult mesozoans develop on the inner surface of the cephalopod's kidney, where they alternate sexual and asexual generations. Within twenty to thirty jacket cells, the hermaphroditic, cylindrical gonad produces both eggs and sperm. Sperm fertilize eggs within the adult, where the tiny zygotes develop into ciliated larvae (**C**), called infusoriform because they resemble ciliates (see page 94). The larvae escape into the ocean in the urine of the mollusc. The free-swimming mesozoan larvae find young cephalopods on the sea bottom and enter their bodies.

Mesozoans also reproduce without sexuality: Inside the larger animal, mesozoans release a second sort of larva—a wormlike (vermiform) type (**D**)—which also swims in urine. Asexually reproduced vermiform larvae stay inside the kidneys of the mollusc.

Phylum: Priapulida. Priapulids are plump, spiny, warty worms that burrow in coral sand and mud by anchoring their anterior and posterior ends alternately as they go. Both mouth and spine-studded proboscis evert (roll inside out, then inside in), allowing priapulids to seize annelids (see page 136) and other priapulids whole. Projections thought to be sensory organs cover the trunk of the priapulid (as shown here with *Meiopriapulus* sp., **E**). From time to time adult priapulids molt, shedding their covering, the chitinous cuticle.

When polychaete annelids with jaws evolved during the Ordovician period, some 440 to 500 million years before the present, polychaetes displaced priapulids from their position as important carnivores of the oceans. Fossils from the Burgess Shale in western Canada disclose evidence that earlier, during the middle Cambrian period (approximately 530 million years ago), the seas were rich in priapulid life.

Phylum: Pentastoma. Pentastome worms are named for the five projections at the anterior ends of their bodies (as shown here on *Armillifera* sp., **F**). Four of these projections are unjointed legs with retractible claws, similar to the legs of tardigrades (see page 134) and onychophorans (see page 144); the fifth protuberance bears the mouth. Some pentastome adults live in the nostrils of snakes and metamorphose through up to three distinct stages—egg, larva, and adult.

Adults mate within the vertebrate that they inhabit, such as the black-tailed gull, *Larus crassirostris* (**G**). The fertilized eggs are deposited along with the feces of the vertebrate. If the eggs land on a plant, they may be eaten along with the plant by an herbivore, such as a rabbit or sheep. In the herbivore's stomach, wormlike larvae emerge from the eggs. If a carnivore eats the herbivore, the spiny pentastome larvae crawl from the mouth or stomach of the carnivore into its nasal passages or lungs. Here the tiny pentastomes embed themselves and are nourished by blood and mucus.

Christie Lyons

Sandy Seashore

Phylum: Platyhelminthes. The worm depicted here is an example of the simplest animals that have organs, the platyhelminth worms, or flatworms. (In Greek, *platy* means "flat," and *helmis* means "worm.") This free-living, marine, carnivorous flatworm (**A**) can evert its pharynx through its mouth; thus, a single opening serves as both mouth and anus for the flatworm's dead-end gut. All flatworms are bilateral (the left half of the body mirrors the right). Their bodies have three layers of tissue but are solid throughout because they lack a circulatory system and a body cavity. Food and oxygen simply diffuse through their ribbonlike bodies.

The tapeworms (class Cestoda) and the flukes (class Trematoda) are the most familiar flatworms. Cestodes absorb nutrients across their body surface. Although they have a gut, trematodes live inside of other animals. Trematodes, such as *Cryptochyle,* are common in seagulls, snails, and fish. They form life history stages appropriate to each animal in which they live. Schistosomes are liver flukes that are symbiotrophs on people in wet tropical areas, such as sub-Saharan Africa and the Caribbean. Many inhabit the livers, muscle tissue, and other parts of the vertebrate body, in which they mate and carry out a complex life history.

Phylum: Echinodermata. Starfish are the most familiar echinoderms (from the Greek for "spiny skin"), but sea stars, sea biscuits, sand dollars, sea cucumbers, sea lilies, and brittle stars are also members of this phylum. Hard spines and tiny protective organs with pincers stud the surfaces of most echinoderms, including the daisy serpent star, *Ophiopholis aculeata* (**B**). Near the center of the arms is a porous sieve plate called a madreporite, through which the brittle star pulls sea water. Sea water carries oxygen and fills the star's hydraulic system (called the water vascular system), which is a ring canal connected to tube feet with hydraulic suckers.

Echinoderms range the ocean floors from the Arctic Circle south to Antarctica and include sea stars; predators on oysters, mussels, and clams; basket stars with many arms; and the garlic-scented leather stars. Echinoderms are believed to have evolved from mobile, bilaterally symmetrical ancestors that became more sessile and radial, like present-day sea lilies and feather stars.

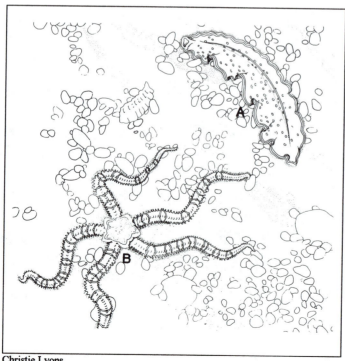

Christie Lyons

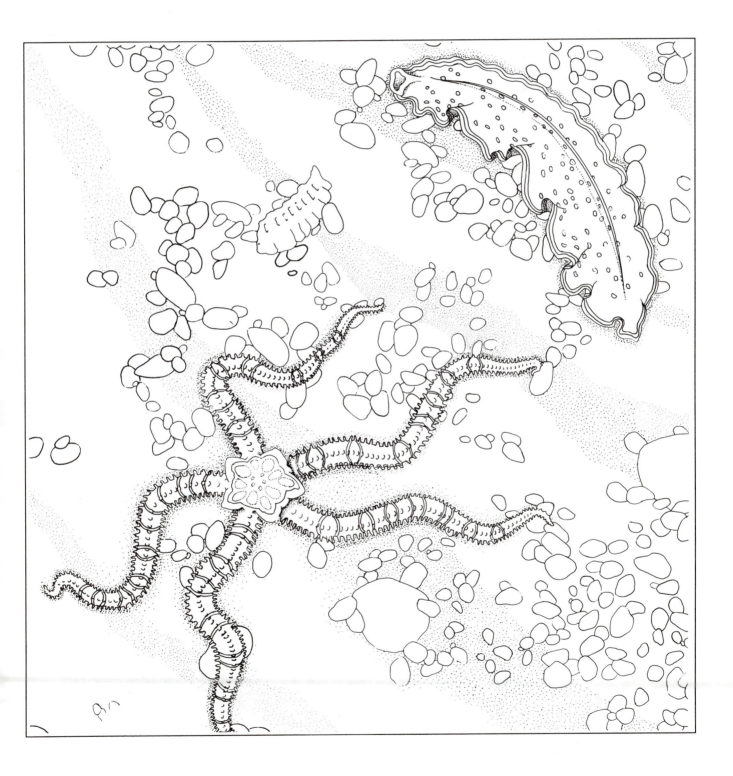

Coral and Squirrel Fish

Phylum: Nemertina. Often very colorful, nemertines such as this ribbon worm (**A**) clambering over the coral *Eunicella verrucosa* (**B**), are mostly marine animals. Some ribbon worms shoot out a hook mounted on a proboscis to capture live prey and then retract the harpooned prey into their mouth. This elaborate proboscis is characteristic of the delicate ribbon worms. The one-way digestive tract of ribbon worms, which have a separate mouth and anus, allows gut regions to be specialized for various steps of digestion and therefore is more efficient than the dead-end digestive tube of flatworms (see page 128).

Nonsegmented ribbon worms differ from other sea worms (such as marine annelids, see page 136), which are segmented and usually bear bristles and paddle feet or parapodia, from tube-dwelling, trophosome-bearing vestimentiferans (see page 146), and from the much smaller gnathostomulids (see page 132).

The squirrel fish, *Holocentrus xantherythrus* (**C**), is one of myriad reef fish species (see Chordata, page 142). Coral gardens carpeting warm, shallow seas around the globe nourish fishes and sea turtles in their plankton- and oxygen-rich waters. Within the elegant branches of corals, tiny marine animals find shelter. Coral polyps secrete minerals that make up coral skeletons (see page 8); masses of corals, along with other animals, monerans, and protoctists, are the reef community.

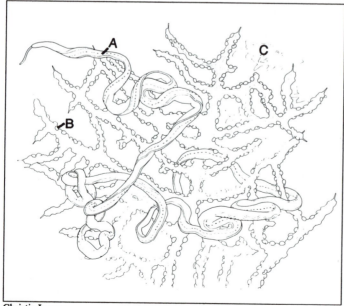

Christie Lyons

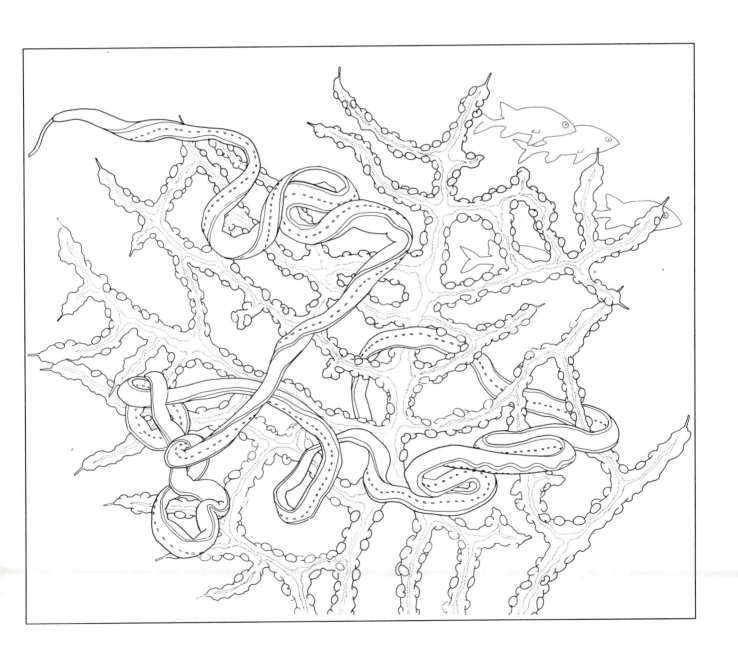

Phylum: Gnathostomulida. All of the gnathostomulids (from the Greek words for "jaw" and "mouth"), or jaw worms, have hard, toothed jaws with which they tear apart fungi and graze on bacteria and protoctists that coat black ocean sands. These transparent microscopic jaw worms inhabit the sulfuretum, an environment in which sand smells like rotten eggs. The smell comes from hydrogen sulfide produced by marine bacteria, under anoxic conditions (i.e., where oxygen gas is absent).

Gnathostomulids such as *Problognathia minima* (shown here on eelgrass, *Zostera marina*, **A**) glide over marine algae using their external cilia, as the flatworms do (see page 128). Stiff bristles on the head and cilia-lined pits are the sense organs of jaw worms. When samples of black ocean sands are investigated, jaw worms are almost the last animals to emerge from the sand, suggesting that they are highly tolerant of anoxic environments. The details of their anaerobic metabolism, however, are obscure.

Phylum: Gastrotricha. The cilia that cover the ventral surfaces of gastrotrichs are what give this phylum its name, which comes from the Greek for "hairy stomach." Scales, bristles, and spines adorn the sides and backs of many species. Careful search with a hand lens may disclose these wormlike animals clinging to water lily pads and bog moss by the adhesive tubes on their posterior ends and sides.

Marine gastrotrichs that glide among coral and in shallow-water sea sands, such as *Lepidodermella* (**B**), have evolved strategies that maximize reproductive success. They are hermaphrodites: Individual gastrotrichs produce both sperm and egg, but at different times. Most produce their sperm earlier than they do eggs. Freshwater gastrotrichs of most species lack males altogether. The unfertilized eggs of females develop, without benefit of male sperm, into the adult organism.

Two egg types are laid by freshwater gastrotrichs. One type must dry, freeze, or be heated before it cleaves to produce a young gastrotrich. The other is enclosed within a thin wall and divides immediately upon being laid. These reproductive strategies permit gastrotrichs to be more likely to survive harsh and changeable environments.

Phylum: Brachiopoda. A pair of calcium- or phosphorus-containing shells enclose soft mantle tissue, giving brachiopods a superficial resemblance to clams. Moreover, like clams, brachiopods are suspension feeders: Their cilia whip plankton-bearing water into their mouths. The resemblance ends there, however. Brachiopods trap plankton on a crown of hollow, ciliated tentacles stiffened by a calcareous loop. This tentacular crown, called a lophophore, graces species of marine animals from three other phyla as well: Phoronida, Ectoprocta, and Entoprocta (see page 138). Sea water is circulated within the cavity between shells, facilitating respiration and excretion.

These marine lamp shells, as brachiopods such as *Terebratulina septentrionalis* (**C**) are commonly called, had a glorious past: More than 30,000 fossil species have been described. Since the Paleozoic era, however, molluscs (see page 142) and other sea-dwelling organisms from more recently evolved phyla have displaced brachiopods from their ecological niches.

Christie Lyons

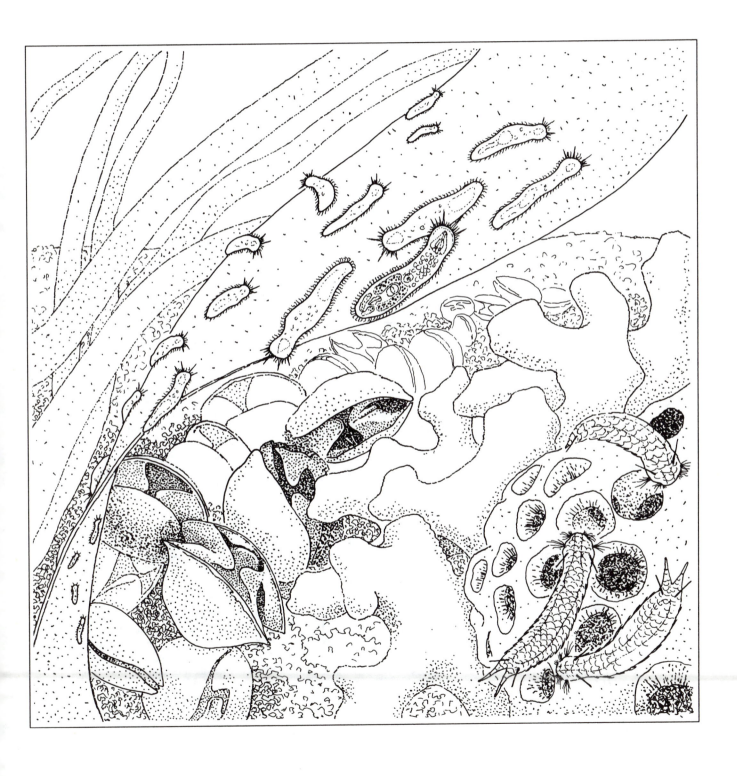

Phylum: Rotifera. Rotifers are small, translucent, cosmopolitan, and aquatic. Many common rotifers (**A**) (e.g., *Philodina*) bear a ciliated crown, which allows them to swim and wafts food to their mouths by causing a current. Because the crown looks like a revolving wheel, rotifers used to be called wheel animalcules. Like tardigrades (see this page), when challenged, rotifers endure adverse environmental conditions. They can reduce their metabolic rate enough to survive subarctic winters and seasonal desiccation. In this condition, they are said to be cryptobiotic. Rotifers and tardigrades share another trait: cell constancy (i.e., the number of cells or cell nuclei is constant).

A transparent shell, called a lorica, encloses most rotifers. Because of their intriguing resistance to freezing and desiccation, rotifers can inhabit freshwater, brackish, or marine habitats—from hot springs and Antarctic lakes to moss, lichens, and tree bark. In a few species, the anterior end is a bristly, lobed funnel that traps prey. Rotifers may feed on other rotifers, bacteria, protoctists, and suspended organic material. In turn, rotifers are food for other aquatic animals, such as carnivorous tardigrades and protists (e.g., the ciliate *Stentor*, see page 94).

Phylum: Nematoda. Of all the animal phyla, the phylum Nematoda contains the most individuals; probably many species are not yet even known. If every nematode

were visible but large organisms were invisible, we would still be able to make out the outlines of plants and animals just by the nematodes living within each of them. Free-living nematodes (**B,** feeding on detritus) abound in soil and are often associated with rotifers and tardigrades.

Predatory nematodes with formidable teeth devour tardigrades and rotifers (see this page), as well as small annelids (see page 136). Nematodes of other species live necrotrophically in plants and animals. Canine and feline heartworm, as well as certain diseases of humans (including river blindness, filarial elephantiasis, trichinosis, hookworm, and pinworm), are associated with nematode infestations.

Nematodes differ from other worms in their unique method of swimming: They generate thrust in S-shaped waves by contracting longitudinal body muscles first on one side and then on the other. Nematodes lack the muscles encircling the body that are characteristic of other worms, such as annelids. The reversible proboscis of ribbon worms (see page 130) and some flatworms (see page 128) is also lacking in nematodes.

Phylum: Tardigrada. In 1772 the Italian biologist Lazzaro Spallanzani used the name *il tardigrado* (meaning "slow walker") for the tardigrades because of their gait. Their common name, water bears, on the other hand, is derived from the habitat in which they live (i.e., the water that covers lichens, moss, and other terrestrial plants and that permeates beach sand). Chubby little animals, tardigrades such as *Echiniscus arctomys* (**C**) typically are about as long as the period that ends this sentence, half a millimeter.

Newly emerging tardigrades pierce the shell that surrounds them with sharp stylets, which protrude and retract through their mouth, and then crawl forth as miniature adults. They feed on the liquid contents of plant or animal cells through holes torn with their stylets or eat nematodes or rotifers (see this page). Claws adorn the tip of each of the tardigrade's four pairs of legs, which lack the joints that are characteristic of arthropod legs (see page 144).

A desiccated tardigrade contracts into a tun (a structure shaped like a wine cask that looks nothing like the adult animal and has less than one-tenth the water of the adult). Tardigrades may remain inert (as tuns) for decades until a water drop revives them. This cryptobiotic behavior may be a clue relating nematodes and rotifers to tardigrades.

Christie Lyons

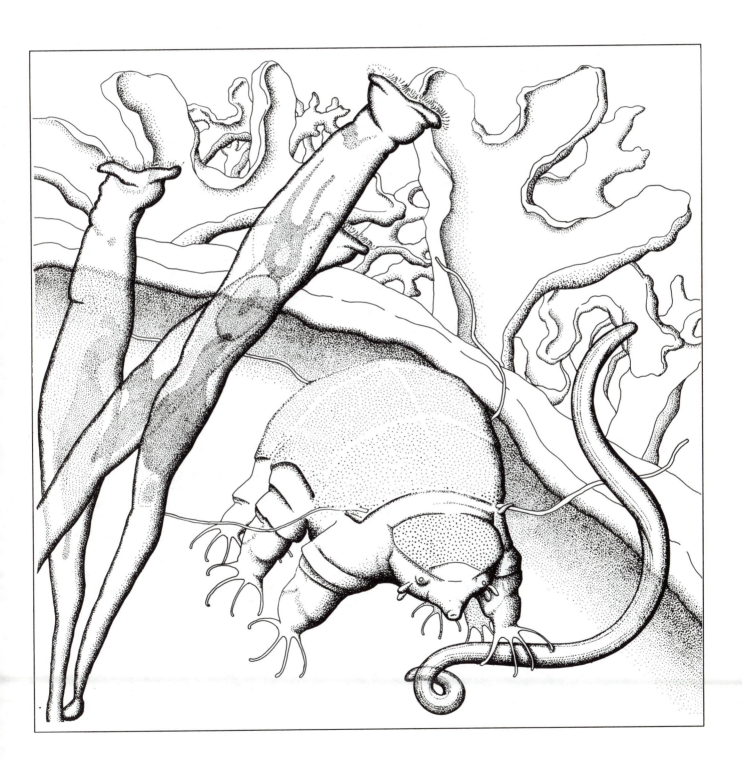

Forest Soil Puddle

Phylum: Acanthocephala. These spiny-headed worms lack any free-living stage; only their cysts exist in the soil. They live as symbiotrophs, at least part of the time in the guts of vertebrate animals—including dolphins, seals, hyenas, moles, shrews, squirrels, pigs, rats, insectivores, turtles, snakes, amphibians, fish, and birds.

This *Plagiorhynchus cylindraceus* larva (**A**) has burrowed into a pill bug (*Armadillium vulgare,* **B**), using the spiny head that characterizes acanthocephalan worms. Eventually, a songbird such as the myrtle warbler, *Dendroica coronata* (**C**), consumes the pill bug along with the worm larva. When eaten by the bird, the acanthocephalan is inside a dark, hard-walled sphere called a cyst. As the bird digests its food, including the *Plagiorhynchus* cysts, the young acanthocephalans are released into the bird's gut, where they absorb nourishment through their body wall, gripping the intestine of the bird with the hooks on their retractable proboscis.

Mating and embryonic development take place within the female worm while it is still inside the warbler. The emerging larvae are released into the bird's gut and exit the bird's anus with its feces. Eventually the larvae are eaten by an arthropod (see page 144) or mammal (see page 142), thus completing the life history.

Phylum: Nematomorpha. King Gordius of the ancient kingdom of Phrygia made an elaborate knot for which the wiry worms of this phylum—Gordian worms—are named. According to legend, King Gordius declared that whoever untied the knot would rule Asia. Alexander the Great sliced the Gordian knot with his sword and annexed Asia to his empire. Gordian worms coil like a Gordian knot in the shallow oceans and freshwater lakes in which they live. Because they are only about one millimeter in diameter but can be as long as a meter, nematomorphs are commonly called horsehair worms.

Gordius villoti (**D**) lays white egg masses. The eggs hatch into larvae, which live symbiotrophically or necrotrophically in grasshoppers, beetles, or cockroaches after being ingested by the insect or after penetrating its epidermis with their spiny front ends (proboscises). Nourished entirely by the animals that they invade, Gordian worms develop into long, thin adults, which are released from the arthropod (see page 144) as eggs either through its intestines or through its decaying body when it dies. These horsehair worm adults hatch into the aquatic world, where they complete their life histories.

Phylum: Annelida. Of all worms, annelids are the most familiar, recognizable by the rings from which their name is derived (in Latin, *annellus* means "little ring"). They include bristleworms (class Polychaeta), earthworms (class Oligochaeta), and leeches (class Hirudinea). All annelids are soft-bodied worms arranged in muscular segments linked by a ventral nervous system. Their fluid-filled coelomic cavity encloses a vascular system and digestive tract. Polychaetes, or marine bristleworms, range from the feather-duster worm with its elegant crown of tentacles to the honeycomb worm, *Sabellaria,* which builds its shell and sand tubes in reefs on rocky coasts of Central America. Bundles of chitinous bristles (chaetae) grow on the paddle feet of polychaetes. Leeches, which have lost their bristles, swim like waterborne caterpillars or creep with the aid of suckers over their amphibian, reptilian, or other prey.

In his book *Formation of Vegetable Mould through the Action of Worms* (1881), Charles Darwin recognized the soil-forming benefits of oligochaetes. Earth would not support the growth of food plants in the absence of earthworms, which aerate the soil with their burrowing and fertilize it with their body waste. Oligochaetes such as the earthworm *Lumbricus terrestris* (**E**) and several other annelids of ponds and mud bear scanty bristles but lack paddle feet, sensory antennae, and the distinctive head (which bears the sensory antennae) of polychaetes.

Christie Lyons

Phylum: Entoprocta. The name entoproct is derived from the Greek words for "inside anus," which is the defining characteristic of this phylum: Both mouth and anus open inside the tentacle crescent. Tiny transparent entoprocts are marine animals on stalks that support a cup-shaped, ciliated calyx. Colonial, sessile entoprocts such as *Barentsia* sp. (**A**) make up animal mats on seaweeds, rock, shell, and other animals in shallow waters. A horizontal stolon links individuals.

Although they reproduce through the fertilization of eggs by sperm, entoprocts, as well as ectoprocts and phoronids (see this page), also reproduce by budding. In unfavorable environmental conditions some entoprocts shed their calyx, then regenerate it when conditions return to normal. Some scientists believe that entoprocts and ectoprocts not only are derived from a common ancestor but also should be united into a single phylum.

Phylum: Ectoprocta. Ectoprocts have a mouth surrounded by tentacles on a reduced head, a U-shaped gut, and ciliated larvae. In these ways the ectoprocts, phoronids, and entoprocts depicted here resemble one another. What distinguishes ectoprocts from phoronids and entoprocts and defines the phylum is that their anus opens outside the crown of tentacles. (The term ectoproct comes from the Latin for "outside anus.")

Colonial ectoprocts thickly encrust marine shells, ship hulls, ocean rock, kelp, and other algae, such as *Sargassum*. Marine fish, such as the rock fish, *Sebastes serriceps* (**B**), nibble bits of these soft sea ectoprocts. Freshwater ectoprocts form gel balls on branches that have fallen into lakes. The soft, living ectoproct individual secretes a nonliving shelter either of the tough, nitrogen-containing polymer called chitin or of a rigid calcium carbonate skeleton overlaid with chitin, within which it is anchored. The shelter is called a zooecium (literally, "animal house," from the Greek *zoo,* meaning "animal," and *oikos,* meaning "home").

The few species of freshwater ectoprocts survive dry and cold conditions by releasing armored balls of cells. Called statoblasts, these living balls begin new colonies of ectoprocts (like the colony of *Bugula* sp. shown here, **C**) as the weather warms and the rivers begin to flow again in the spring. Although calcified boxes within which ectoprocts shelter themselves each may be only about one millimeter across, they make up massive arrays; such marine ectoprocts are important reef formers, along with corals (see page 122) and calcareous marine algae.

Phylum: Phoronida. Phoronid worms live in blind tubes (tubes with only one opening), which they secrete, strengthening their walls with bits of seashell and sand. A structure they have in common with some other animal phyla (e.g., Brachiopoda, see page 132), the lophophore, is fully expanded on the anterior end of *Phoronopsis vancouverensis* (**D**). The lophophore draws detritus and plankton toward the mouth, which opens within the lophophore. The phoronid can quickly snap into its tube, which is about twice the length of the worm, using muscles coordinated by a giant nerve fiber. The U-shaped gut and sheltering tube are adaptations to sessile life.

Adults of some phoronid spescies brood their fertilized eggs between their tentacles, where the ciliated larvae develop. Larvae swim free, disperse, and encrust pilings, rocks, shells, and ocean bottoms. Leathery or chitinous phoronid tubes can be seen protruding from soft sands or entwined like vermicelli.

Christie Lyons

Sandy Marine Coast

Phylum: Sipuncula. Sipunculans, or peanut worms, are marine and sedentary. Bushy, mucus-covered tentacles encircle the mouth of many sipunculans, collecting suspended food particles. Like echiurans (see this page) and priapulids (see page 126), sipunculans extend their introvert—an anterior body part that can turn inside out to feed and completely withdraw into the trunk.

Like the hemichordates, such as *Saccoglossus kowalevskii* (**A**), many sipunculans form burrows. Some, such as *Themiste lageniformis* (**B**), which nestle among corals (see page 122), extend their contractile introvert over the reef surface, scraping off films of algae (see page 84), which they ingest. Sipunculans of shallow seas engulf sand as they burrow along; they rework marine sediments just as earthworms (see page 136) rework soil as they burrow through it eating the detritus.

Organic matter dissolved in ocean water may provide up to 10% of the diet of sipunculans. Although not very abundant, sipunculans are widely distributed. They have been seen in the abyss at 7,000 meters deep in the sea, as well as in icy polar oceans, but most peanut worms reside in warm seas between tidemarks, among mangrove stilt roots and eelgrass, in empty shells, in annelid tubes, or under rocks. Some even live in the sea mouse, *Aphrodite* (phylum Annelida).

Phylum: Echiura. Echiurans, also called spoon worms, are sedentary and marine. An echiuran thrusts forth its grooved, motile, ciliated proboscis to gather detritus as food from its U-shaped burrow. *Urechis caupo* (literally, "the innkeeper"), an echiuran up to 50 cm long, spins a net within its burrow. This mucus web strains minute particles, as small as 0.04 micrometer wide, from ocean water. Every several minutes, the innkeeper swallows its food-coated net, then secretes a fresh net.

Echiurans pump sea water through their burrow by alternately relaxing and constricting their trunk muscles; the currents of water oxygenate their blood, move food through the tunnel, and remove waste. *Urechis* is a suspension feeder; most other echiurans, such as *Listriolobus pellodes* (**C**), are deposit feeders, taking their food (bacteria, algae, protoctists, and plant and animal parts) off of the ocean floor. The comma-shaped impressions (**D**) are feeding traces made by the echiuran as it sweeps its proboscis over the ocean floor in search of food.

Echiuran and sipunculan larvae greatly resemble the ciliated, free-swimming larvae of the hemichordate (see page 120) *Balanoglossus* (**E**).

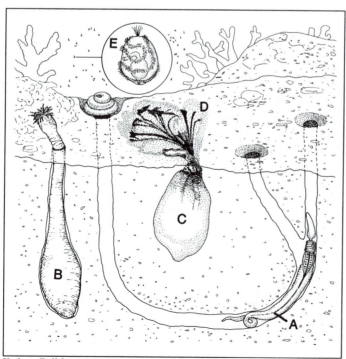

Kathryn Delisle

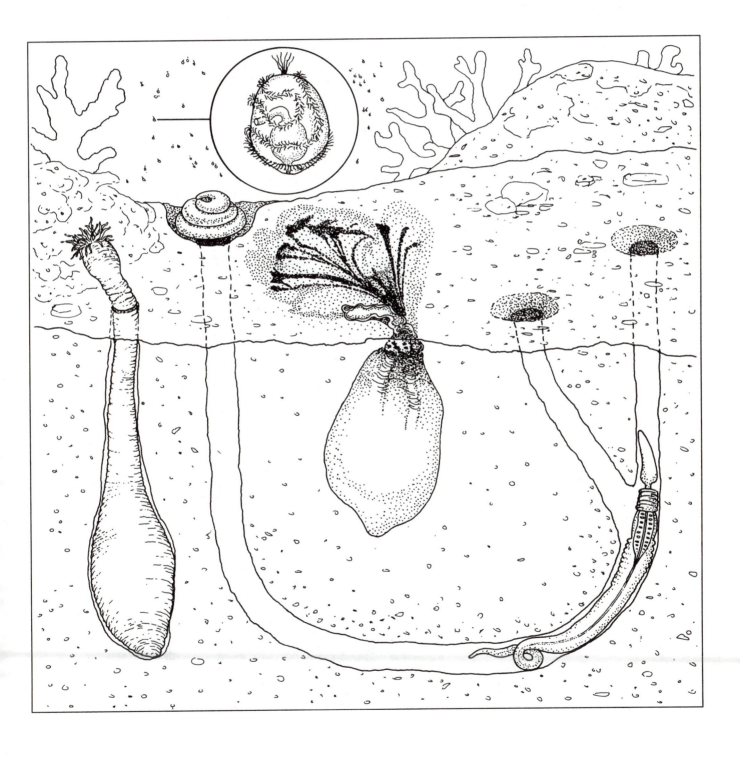

California Rocky Intertidal Zone

Phylum: Mollusca. The familiar molluscs include scallops, snails, squid, and octopi. "Shellfish" (which is not a biological term) refers to many different molluscs, including clams, oysters, and mussels. Molluscs are soft-bodied animals; most have an internal or external shell. All molluscs also have a mantle, a fold in the body wall that lines the shell and secretes the calcium carbonate of which the shell is made. Unique to molluscs is the radula—a hard, chitinous strap that bores or scapes like a file and is used to gather food.

Molluscs live in the water or in moist land environments. They are well-known inhabitants of mud and sandy flats, and they also dwell in forests, soil, rivers, lakes, and the abyss of the sea. Molluscs are divided into several groups, including gastropods, cephalopods, chitons, and bivalves.

The gastropods include the black abalone, *Haliotis cracherodii* (**A,** shown here being eaten by a sea otter), and the naked shell-less slug, the purple nudibranch, *Flabellinopsis iodinea* (**B**). Cephalopods include the common squid, *Loligo pealei* (**C**), and the Pacific giant octopus, *Octopus dofleini* (**D**). All bivalves, such as the California mussel, *Mytilus californianus* (**E**), are members of the class Pelecypoda. Chitons, such as *Stenoplax* sp. (**F**), belong to the class Polyplacophora; they have distinctive external shells of eight overlapping plates embedded in their mantle.

The octopus and squid, both cephalopods, show the head and mantle of molluscs most clearly. A calcareous shell shelters the soft body of more sedentary molluscs, such as the abalone cradled on the chest of the otter (**A**). Descending into the sea, chitons, mussels, and limpets (**G**) cling to rocks. Although at first glance the barnacles (*Balanus,* **H**) appear to be shelled molluscs, closer inspection reveals that their legs are jointed, making them arthropods (see page 144). A forest of *Laminaria,* a phaeophyte (see page 62), carpets this subtidal zone of the California sea otter's world.

Phylum: Chordata. The chordates include mammals, birds, reptiles, amphibians, and fish (i.e., all of the most familiar animals), as well as many organisms that are known only to professional zoologists. All chordates, such as the sea otter (*Enhydra lutris,* **J**), have a tail, a pharynx with gill slits, a notochord, and a hollow dorsal nerve cord at some time during their life history. (The term chordate comes from the Latin word *chorda,* meaning "cord.")

The gill slits of otters and other mammals close during embryonic development; fish and some amphibians retain permanent gill slits. The notochord, a flexible support rod that runs from head to tail, is permanent in adult lancelets (subphylum Cephalochordata), fishlike chordates that lack bones and cartilage. Most chordates belong to the subphylum Vertebrata. In vertebrate animals, including otters and people, a segmented vertebral column—the familiar neck and back bones—replaces the notochord.

We humans (members of the genus *Homo* and species *sapiens*) have notochords only before birth in the fetal stages of life. Sharks, rays, skates, lampreys, and hagfish are supported by a vertebral column of cartilage, whereas bony fish, amphibians, reptiles, birds, and mammals have bony vertebrae. The dorsal hollow nerve cord that is also characteristic of chordates is the spinal cord of humans. Nonchordate animals with well-developed nervous systems differ distinctly: They have solid nerve cords on the belly, or ventral side, of their bodies instead of on the dorsal side.

Christie Lyons

Costa Rican Tropical Forest at Night

Phylum: Onychophora. Onychophorans, or velvet worms (like *Peripatus* sp., **A**), venture forth in the night to feed. Carnivorous onychophorans hunt wood lice and terrestrial molluscs in wet, warm woods like this Costa Rican forest. Champion spitters, these velvety animals rear up and spit whitish fluid 80 to 300 millimeters (much further than their body length of 14 to 150 millimeters). Adhesive glands secrete the fluid, which gels in contact with air, tangling prey in a sticky net. An onychophoran bites into trapped prey, then secretes into the prey substances that kill it and partially liquefy its tissues. Some onychophorans feed on dead termites and the carcasses of larger insects, such as grasshoppers. Food habits of species in this phylum are carnivorous, herbivorous, or omnivorous.

Velvet worms have hooked claws on seventeen to forty-three pairs of unjointed, hollow legs; soft bodies; and a pair of delicate, ringed antennae—a unique combination of annelid (see page 136) and arthropod (see this page) characteristics. Like annelids, onychophorans have body-wall muscles that are smooth rather than striated, a single pair of jaws, unjointed appendages, and

locomotion based on a hydrostatic skeleton. Their body segments have a pair of nephridia (kidneylike excretory organs). The simple eyes (ocelli) of onychophorans sense light. Even professional zoologists have difficulty finding the elusive fossils because velvet worms have no bones. Evidence from ribosomal RNA sequences suggests that onychophorans may be modified arthropods.

Phylum: Arthropoda. Arthropods (from the Latin for "joint-footed") include insects, such as this saturniid moth (*Rothschildia lebeau,* **B**), as well as crustaceans, centipedes, and spiders. *R. lebeau* was named after Walter Rothschild (1868-1937), an avid British collector of museum specimens who classified more than two million insects himself. The golden beetle (*Plusiotis aurigans,* **C**), like all other arthropods, is distinguished by its jointed antennae and legs. In contrast to the soft, flexible cuticle of onychophorans (see this page), arthropods secrete a stiff, jointed exoskeleton.

Arthropods and onychophorans share some characteristics, however: Their cuticle contains chitin; the body cavity (hemocoel) is filled with fluid and is part of the circulatory system; and their tubular heart has slitlike openings called ostia. Another link between these two phyla is that the adhesive defense and feeding fluids of millipedes and centipedes, which are arthropods, resemble those of onychophorans.

In number of species, the phylum Arthropoda is by far the largest in the animal kingdom. Nearly half a million species of insects alone have been described. Some zoologists believe that, if the tropical groups were known better, as many as ten million living species of insects would be recognized.

Christie Lyons

Marine Abyss

Phylum: Vestimentifera. This dark community of tube worms, crabs, and clams functions far below the sun-driven food webs on the thin skin of Earth. In these sub-marine gardens, sea vents spew forth steam and hot water as much as 2.5 kilometers (1.5 miles) below the ocean surface. Vestimentiferans gain life-supporting energy via symbiotic bacteria that oxidize sulfide and methane in the sea water. (The oxygen involved originates at Earth's surface.) Tube worms such as *Riftia pachyptila* (**E**) and their close relatives, the pogonophorans, may be nourished by molecular food absorbed through the plume of tentacles that projects from their tough tubes, but mostly they depend on symbiosis with the bacteria that pack their midgut for their food supply.

From the tiniest (0.75 millimeters long) to the giant (1.5 meter long) tube worms, neither gut nor mouth is present, except in juveniles. Free-living bacteria may pass through the transient mouth of the young worm and later reproduce and pack the midgut (trophosome) of the tube worm. Animals of most other phyla derive energy and nutrients such as nitrogen, carbon, and phosphorus from organic compounds; such organisms are heterotrophs. By contrast, tube worms harbor chemoautotrophic bacteria (see page 42) within their bodies and are autotrophs (literally, "self-feeders")—i.e., organisms that grow and synthesize organic compounds by oxidizing inorganic compounds, such as sulfur.

Three billion years in the past, hydrothermal vents belching forth sulfides and methane were probably typical environments for vestimentiferans. (See page 66 for identification of **A, B, C,** and **D**.)

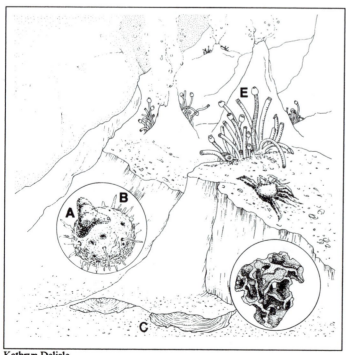

Kathryn Delisle

Chapter 5: Plantae

All members of the kingdom Plantae develop from embryos, an early developmental stage of multicellular organisms produced from a fertilized egg (Figure 14). (Of all the large eukaryotes, only fungi never form embryos.) Because plants develop from embryos, they are multicellular. Furthermore, since embryos are the products of the sexual fusion of female and male nuclei, all plants potentially (although perhaps not in reality) have a sexual stage in their life cycle. Most often this sexual stage involves the male cell (sperm or pollen nucleus) fertilizing the female (egg or embryo-sac nucleus). Many plants are capable of growth and reproduction in ways that bypass two-parent sexual fusion, but all plants evolved from ancestors that formed embryos by sexual cell fusion.

Composed of nucleated cells, plants are eukaryotes. Plant cells are enclosed within cellulosic walls and have green plastids (chloroplasts), which contain pigments such as chlorophylls *a* and *b*, xanthophylls, and other yellow and red carotenoids. Plant cells tend to be larger than those of animals and to have conspicuous internal flow, which is called cytoplasmic streaming, or cyclosis. Plants are the major transformers of solar energy into food, fiber, fuel, and pharmaceuticals. We use plant products in myriad forms. Photosynthesis by plants sustains the entire biosphere, not only converting solar energy into food, but also producing oxygen.

All 500,000 species of plants belong to one of two basic groups: the bryophytes (nonvascular plants) or the tracheophytes (vascular plants). Familiar woody and herbaceous plants tend to be tracheophytes, which are distinguished from the nonvascular plants by conducting tissues called xylem and phloem. The xylem transports water and ions from the roots upward through the plant, and the phloem transports photosynthate, sugar, and other products of the leaves throughout the plant. Like those of their green algal ancestors (the Chlorophyta in the kingdom Protoctista, see page 84), the cells of green land plants often are interconnected by plasmodesmata. These cell-to-cell connections link the walls of adjacent cells and are sufficiently large macromolecules such as sugar to pass through via cytoplasmic strands. Mosses, liverworts, and hornworts—the only bryophytes living today—lack these rigid but metabolically active tissues. Bryophytes can be found mostly in swampy places, on the rocks of waterfalls, and at river's edge in humid climates.

Kathryn Delisle

Figure 14
Plant reproductive structures. **Spores:** *Equisetum* sp., horsetail, with elaters open (1) and coiled (2); microspore of *Ginkgo biloba,* and empty tetrad walls (3); *Sphaerocarpos* sp., liverwort (4). **Pollen grains:** *Cometes surattenis* (5); *Pelucha trifida* (6); *Cerastium alpinum* (7). Horseshoe-shaped **embryo** (arrow) within seed (*Capsella bursa-pastoris,* shepherd's haircap) (8). **Sperm:** *Zamia* sp., coontie (9); *Polytrichum juniperinum,* moss (10). **Seeds:** generalized dicot (11) and monocot (12) seeds; winged seed (13) and cross section of *Pinus* sp. (14).

Woods

Phylum: Bryophyta. The most common and well-known bryophytes (low-lying, leafy, nonvascular plants) are the mosses (class Musci), such as *Polytrichum juniperinum* (**A**). In chromosomal composition and dependency of their sperm on water, these nonvascular plants bear the closest resemblance of all plants to their protoctist chlorophyte ancestors (see page 84). Although the form of angiosperms (see page 162) and other vascular plants that we are most likely to recognize is the diploid sporophyte generation, the most familiar form of bryophytes is the haploid gametophyte generation.

Sexual reproduction in bryophytes requires water so that the sperm (each with two forward-directed undulipodia) can travel from the antheridia (the male sex organs) to the archegonia (the female sex organs) for fertilization. The archegonia usually are swellings, on leafy surfaces, that harbor unfertilized egg cells. After egg and sperm fuse, the zygote forms in the archegonium. From the zygote arises the sporophyte (**B**), with its slender stalk (seta) and spore capsule. Spores are formed by meiosis at the tip of the sporophyte.

Mosses—for example, the *Sphagnum* moss of peat bogs—are common throughout the world. The other, less common members of this phylum are the liverworts (class Hepaticae) and the hornworts (class Anthocerotae).

Phylum: Lycopodophyta. Some lycopods, known as club mosses or ground pines and represented by the genus *Lycopodium,* can be found throughout the temperate forests of the United States. Others, such as the quillworts (genus *Isoetes*), are far less common. Both of these nonflowering vascular plants, however, display distinctive sterile microphylls—small evergreen leaves that probably evolved as outgrowths of the main photosynthetic axis.

The haploid gametophytes grow from haploid spores and are often subterranean, living in symbiosis with mycorrhizal fungi. The underground rhizome of *Lycopodium obscurum* (**C**) connects the aboveground, green, herbaceous sporophytes. Some species of lycopods have their sporophylls interspersed among the small leaves. Atop the sporophytes of other lycopods are nonchlorophyllous strobili, which bear the spore-forming organs, the sporangia. These club-shaped strobili give rise to the misnomer "club mosses."

Brown, spore-bearing leaves called sporophylls are the modified leaves that produce sporangia, in which haploid spores develop by meiosis. The spores drop to the soil, where they germinate to form the gametophyte. Although strobili are club-shaped, lycopods are not mosses at all. These plants, whose huge ancestors were the size of trees, many species of which dominated the coal-forming forests of the Carboniferous period, are easily recognizable denizens of temperate forests or live as epiphytes in the tropics.

Phylum: Filicinophyta. Today ferns are the most widespread and well known of the seedless, vascular plants. Ferns have distinctive megaphylls called fronds (**D**)--large, photosynthetic structures that develop spores. Although often called leaves, propagules called fronds differ botanically from the leaves of angiosperms (see page 162). The large, conspicuous sporophyte generation grows from a small, heart-shaped gametophyte, where motile sperm fertilize eggs.

Some fern species, such as the cinnamon fern, *Osmunda cinnamomea* (**E**), produce sporangia on a separate stalk; other species form clusters of sporangia called sori on the undersides of fronds. An underground, persistent rhizome produces a distinctive fiddlehead bud (**F**) in the spring. Many of the 12,000 species of ferns are tropical. Some ferns, such as the tiny *Azolla*, grow on the surface of water.

Kathryn Delisle

Florida Bush

Phylum: Psilophyta. This phylum consists of just two living genera, one of which, *Psilotum* (the whisk fern), is pictured here (**A**). Plants of this genus are unique because they lack both leaves and roots, giving them a resemblance to the earliest vascular plants, the rhyniophytes, although whisk ferns may have evolved from a fernlike ancestor. As with the other seedless vascular plants, psilophytes alternate generations. The sporophyte is the prominent, diploid form, whose spores are shed by small swellings. The haploid spores grow into small, inconspicuous, subterranean plants: the bisexual gametophytes (**B**). Endomycorrhizae, which are fungal connections with subterranean, rootlike projections, are established in both generations. As with *Lycopodium* (see page 152), whisk ferns grow from an underground rhizome (**C**). Dehiscent, homosporous sporangia, borne in groups of three called synangia, produce spores (**D**).

Phylum: Cycadophyta. Cycads traditionally have been classified as gymnosperms because their seeds are naked, not inside an ovary. Some cycads are called sago palms; however, they resemble and are more closely related to flowering plants (see page 162) than to conifers (see page 156). Cycads have compound fern- or palm-like leaves and produce seeds. Approximately one hundred species of cycads exist today, mostly in the tropics and subtropics. *Zamia* (coontie, **E**), common in southern Florida (where native vegetation has not yet been destroyed) and the only cycad native to the continental United States, is pictured here.

Most cycads are dioecious: Male and female cones are found on different plants. A large cone grows on the female plant (**F**); the cone of the male plant is longer and thinner (**G**). Pollen produced by the male cone is blown by wind to the female cone. Sperm are liberated from the germinating pollen grains and bear undulipodia. In the moist female cone the sperm swim to fertilize the egg. Cycads were very abundant and diverse in the Mesozoic period, the time of the dinosaurs.

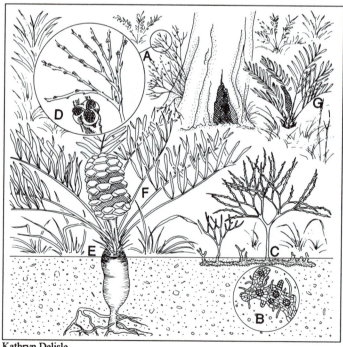

Kathryn Delisle

Pitch Pine Barrens

Phylum: Sphenophyta. Sphenophytes, including the common horsetail, *Equisetum arvense* (**A**), are easily recognized by their jointed hollow stems and rough, ribbed texture, which is due to hardening caused by silica (SiO_2) embedded in the tissue. (This texture makes the plant good for cleaning pans and gives the plant its nickname, "scouring rushes.") The presence of silica in the tissue indicates that sphenophytes are capable of biomineralization, the production of minerals by cells.

Ancestors of the sphenophytes flourished in the Carboniferous period, when many of their treelike relatives were as tall as fifteen meters and woody. Today only forty species of these seedless vascular plants remain, all smaller than their illustrious ancestors and all members of the genus *Equisetum*. Some live on salt flats, others along the banks of streams, and many others in colonies in wooded areas.

The visible form of the plant is the diploid sporophyte, with the sporangia located in the strobilus, or cone, at the top (**B**). The cones are the site of meiosis, during which diploid cells are reduced to haploid spores, which drop to the soil and develop into gametophytes. Horsetail sporophytes also propagate by underground stems (**C**). As with ferns (see page 152), the sporangia often are borne on stalks separate from those bearing the photosynthetic leaves. *Equisetum arvense* is common in wasteland areas and on silica-rich soils.

Phylum: Coniferophyta. Conifers such as *Pinus rigida* (**D**), pitch pine, are the most familiar of the gymnosperms (plants that produce naked seeds). Conifers are grouped into fifty genera, including *Pinus* (pine), *Taxus* (yew), *Abies* (fir), *Picea* (spruce), and *Larix* (larch). The phylum also includes the largest living plant, *Sequoiadendron gigantea,* the giant sequoia of California.

The seeds of conifers are naked in the sense that the embryos are not covered by the diploid tissue of the ovary wall from the female parent. Rather, embryos are embedded in the megasporangium, the haploid tissue of the parent that produces megaspores. Most of these cone-bearing plants are softwood trees with needles. Some are shrubs or low-lying plants with needles.

Most conifers are also monoecious (male and female reproductive structures are borne on the same plant). Conifers are also heterosporous: Male microsporangia are borne on small, pollen-producing cones (**E**), whereas female megasporangia are borne on the larger, more familiar pine cones (**F**). In contrast to fern sperm (see page 152), which may bear more than 100 undulipodia, the sperm of conifers lack undulipodia altogether.

Kathryn Delisle

Hillside in China

Phylum: Ginkgophyta. Since only one species of this living fossil exists today, we are lucky that it is a hardy ornamental often planted along streets. The only wild stands of ginkgo trees left grow vigorously on the steep hillsides of southern China. Although the fleshy female cone stinks like rotten meat, the seeds inside, when roasted, are considered delicious. If injured at its base, the ginkgo tree forms what the Chinese call *chi-chi's,* outgrowths capable of spectacular healing that restore the tissue so well, it is as if the tissue had not aged at all. From the swollen *chi-chi* (**A**), healthy young tissue sprouts, again and again.

Ginkgo biloba is the only living descendant of a group of trees that, during the Mesozoic period, was far more extensive. Although it is very popular in gardens, on city streets, and on temple grounds all over Asia, where people tend it and even eat its roasted "nuts," in the wild the ginkgo tree is restricted to extremely steep mountain slopes in southern China. Its deciduous, fan-shaped leaves are borne close to the richly branching stems, giving the trees the characteristic silhouette that makes them so easy to recognize, even from a distance.

Ginkgos are dioecious: Male (**B**) and female (**C**) reproductive structures are borne on different plants. Pollen is carried by wind to female trees, where it releases undulipodiated sperm that swim through maternal tissue, ultimately fertilizing the egg. No external water is required because the tissues are so juicy that the sperm can reach the egg just by beating their hundreds of undulipodia. Sperm develop only during a few days in the spring.

Ginkgos are sometimes referred to as maidenhair trees because their leaves resemble the fronds of the maidenhair fern. Ginkgo extract is now one of the most widely used prescription drugs in western Europe for infirmities from asthma to Alzheimer's disease.

Kathryn Delisle

African Desert

Phylum: Gnetophyta. Gnetophytes include a miscellany of cone-bearing, seed-producing plants classified into three perhaps unrelated genera: *Gnetum, Ephedra*, and *Welwitschia*. *Welwitschia mirabilis* (**A,** old plant with tattered leaves; **B,** younger plant; **C,** young plant) is a strange plant that lives only in the deserts of Namibia, in southwestern Africa. To see this remarkable plant alive requires a visit to Namibia or to one of the great botanical gardens of the world (Kew in London, the St. Petersburg Garden, or the New York Botanical Garden).

With its distinctive, straplike leaves attached to a woody, cup-shaped stem, *Welwitschia* literally lies down on the desert sand. The leaves grow continuously for the entire life of the plant--up to a hundred years. The stem is mostly underground and, along with the root that reaches down to the water table, serves as the water-storage facility of the plant. Like conifers (see page 156), *Welwitschia* is monoecious and heterosporous, with female (**D**) and male (**E**) cones, each having distinctive spores, borne on a single plant. The cones of *Welwitschia,* compound and with opposite or whorled bracts, are different from those of conifers.

Species of *Ephedra* produce the vasoconstrictor drug ephedrine, which is used for treating asthma and hay fever and to stop nosebleeds. Growing wild in the deserts of the southwestern United States, *Ephedra* is sometimes called Indian tea because the medicinal value of its infusions is recognized by native Americans.

Kathryn Delisle

Rain Forest Canopy

Phylum: Angiospermophyta. With more than 400,000 species, the angiosperms, or flowering plants, are today the most widespread, diverse, and successful plants. During the past 100 million years, angiosperms have replaced cycads (see page 154), ginkgos (see page 158), and conifers (see page 156) as the dominant phylum of plants on Earth. Divided into some 350 families, such as the Bromeliaceae (**A,** e.g., *Vriesea ringens* or *Tillandsia asplundia*) and the Orchidaceae (**B,** e.g., *Epidendrum* or *Encyclia*), angiosperms have evolved a wide variety of life forms from trees to herbs to vines and epiphytes that grow in the tropical rain forest canopy.

The flower and its ovaries, which develop into fruit that enclose seeds, are the distinctive characteristics of flowering plants. In these flowers, the alternation of generations that is so distinctive in other plants is reduced to microscopic size, with the male microgametophyte, or pollen sac, consisting of just three cells and the female megagametophyte, or embryo sac, made up of just seven cells. Fully capable of life on land, flowering-plant pollen grains grow tubes that convey sperm nuclei to eggs inside the ovaries deep within the flowers.

Insects (see page 144), like the butterfly shown here, are crucial to the survival of flowering plants; many depend on insects, bats, or small mammals for pollination. Other angiosperm pollen is borne by wind. Because of this relationship, insects have co-evolved with angiosperms, have spread with them across the planet, and enjoy an equally diverse and pervasive presence in the animal kingdom.

Kathryn Delisle

Appendix: Classification

Listed here are the phyla illustrated in our five-kingdom classification, the representative genera included in the drawings, and the habitats in which they are found. Common names are given where applicable. Page numbers for the phyla and genera are given in the Index (beginning on page 219). Organisms discussed but not illustrated also may be located with the Index.

PHYLUM (COMMON NAME)	REPRESENTATIVE(S)	HABITAT (PAGE NO.)

Kingdom Monera

PHYLUM (COMMON NAME)	REPRESENTATIVE(S)	HABITAT (PAGE NO.)
Actinobacteria	*Cellulomonas*	Woodland Stream (46)
	Frankia	Woodland Stream (46)
Aeroendospora (aerobic, endospore-forming bacteria)	*Bacillus*	Rocky Brook (30)
Anaerobic Phototrophic Bacteria	*Chromatium*	Cover illustration (iv)
	Rhodomicrobium vannielii	Pond Scum (32)
Aphragmabacteria	*Mycoplasma*	Barnyard (26)
Chemoautotrophic Bacteria	*Nitrobacter winogradskyi*	Ocean Edge (42)
	Nitrosococcus	Ocean Edge (42)
	Nitrosolobus	Ocean Edge (42)
	Nitrosospira	Ocean Edge (42)
	Nitrosovibrio	Ocean Edge (42)
Cyanobacteria	*Anabaena*	Cover illustration (iv)
	Gloeothece	Salt Marsh (34)
	Johannesbaptista	Salt Marsh (34)
	Microcoleus	Salt Marsh (34)
	Oscillatoria limnetica	Salt Marsh (34)
	Prochloron	South Pacific Coral Reef (36)
	Spirulina	Salt Marsh (34)

PHYLUM (COMMON NAME)	REPRESENTATIVE(S)	HABITAT (PAGE NO.)

Kingdom Monera (continued)

Fermenting Bacteria	*Lactobacillus*	Foods (44)
Halophilic and Thermoacidophilic Bacteria	*Halobacter*	Hot Springs and Mud Flats (24)
	Halococcus	Hot Springs and Mud Flats (24)
	Thermoplasma acidophilum	Hot Springs and Mud Flats (24)
Methanocreatrices (methanogens)	*Methanobacterium*	Pasture (22)
	Methanobacterium ruminantium	Pasture (22)
	Methanogenium cariaci	Pasture (22)
	Methanogenium marirgen	Pasture (22)
	Methanosarcina	Pasture (22)
Micrococci	*Micrococcus*	Garden Soil (38)
Myxobacteria	*Stigmatella aurantiaca*	Garden Soil (38)
Nitrogen-fixing Aerobic Bacteria	*Azotobacter*	Garden Soil (38)
Omnibacteria	*Aeromonas punctata*	Lake Shore (40)
	Caulobacter	Lake Shore (40)
	Escherichia coli	Cover illustration (iv) Lake Shore (40)
Pseudomonads	*Pseudomonas multivorans*	Pond Scum (32)
	Xanthomonas	Pond Scum (32)
Spirochaetae (spirochetes)	*Cristispira*	Clam Camp (28)
	leptospire	Clam Camp (28)
	Spirochaeta	Clam Camp (28)
Thiopneutes	*Desulfovibrio*	Rocky Brook (30)

Kingdom Protoctista

Acrasea (acrasids)	*Acrasis rosea*	Birch Forest Floor (60)
	Copromyxella spicata	Birch Forest Floor (60)

PHYLUM (COMMON NAME)	REPRESENTATIVE(S)	HABITAT (PAGE NO.)

Kingdom Protoctista (continued)

Actinopoda (actinopods)	*Challengeron wyvillei* (phaeodarian)	Ocean Water Column (80)
	Clathrulina fragilis (heliozoan)	Ocean Water Column (80)
	Phyllostaurus siculus (acantharian)	Ocean Water Column (80)
	Spongosphaera polyacantha (polycystine)	Ocean Water Column (80)
Apicomplexa (apicomplexans)	*Plasmodium*	Mosquito and Blood (98)
Bacillariophyta (diatoms)	*Amphora*	Oceanside (100)
	Cocconeis	Oceanside (100)
	Fragilaria	Oceanside (100)
	Gomphonema	Oceanside (100)
	Melosira	Oceanside (100)
	Opephora	Oceanside (100)
	Rhoicosphenia	Oceanside (100)
Chlorarachnida (chlorarachnids)	*Chlorarachnion reptans*	River Delta (70)
Chlorophyta (chlorophytes)	*Chaetosiphon moniliformis*	Atlantic Sheltered Bay (84)
	Cladophora	Oceanside (100)
	Oedogonium	Decaying Pond Vegetation (92)
	Ulva lactuca (sea lettuce)	Atlantic Sheltered Bay (84)
Chrysophyta (chrysophytes; golden-yellow algae)	*Dinobryon*	Farm Pond (90)
	Ochromonas	Farm Pond (90)
	Synura	Farm Pond (90)
Chytridiomycota (chytridiomycotes)	*Blastocladiella emersonii*	Decaying Pond Vegetation (92)
	Rhizophydium granulosporum	Decaying Pond Vegetation (92)
	Spizellomyces	Decaying Pond Vegetation (92)
Ciliophora (ciliates)	*Bursaria*	Pond Rocks (94)
	Paramecium bursaria	Pond Rocks (94)
	Stentor	Pond Rocks (94)
	Tetrahymena	Pond Rocks (94)
	Vorticella	Pond Rocks (94)

PHYLUM (COMMON NAME)	REPRESENTATIVE(S)	HABITAT (PAGE NO.)

Kingdom Protoctista (continued)

Conjugaphyta (gamophytes)	*Cosmarium*	Shallow Pond (64)
	Micrasterias denticulata	Shallow Pond (64)
	Spirogyra	Shallow Pond (64)
	Zygnema	Shallow Pond (64)
Cryptophyta (cryptomonads)	*Cyathomonas truncata*	Ice-covered Lake (68)
Dictyostelida (dictyostelids)	*Polysphondylium violaceum*	Birch Forest Floor (60)
Dinomastigota (dinomastigotes)	*Gonyaulax tamarensis*	Coastal Red Tide (88)
	Pyrocystis	Coastal Red Tide (88)
Ebridians	*Ebria tripartita*	Pacific Nearshore Waters (104)
Ellobiopsida (ellobiopsids)	*Thalassomyces marsupii*	Pacific Nearshore Waters (104)
Euglenida (euglenids)	*Euglena gracilis*	Outfall Pipe (74)
	Euglena spirogyra	Outfall Pipe (74)
	Phacus	Outfall Pipe (74)
Eustigmatophyta (eustigmatophytes)	*Vischeria*	Lake Surface (78)
Glaucocystophyta (glaucocystophytes)	*Gloeochaete*	River Delta (70)
Granuloreticulosa (granuloreticulosans)	*Globigerinoides*	Tropical Coast (96)
	Heterotheca lobata	Tropical Coast (96)
	Textularia	Tropical Coast (96)
Haplosporidia (haplosporidians)	*Haplosporidium nelsoni*	Estuary with Oyster Bar (54)
Hyphochytriomycota (hyphochytrids)	*Hyphochytrium catenoides*	Pine Pollen (82)
Karyoblastea (karyoblasteans)	*Pelomyxa palustris*	Swamp (52)

PHYLUM (COMMON NAME)	REPRESENTATIVE(S)	HABITAT (PAGE NO.)

Kingdom Protoctista (continued)

PHYLUM (COMMON NAME)	REPRESENTATIVE(S)	HABITAT (PAGE NO.)
Labyrinthulomycota (labyrinthu-lomycotes; slime nets)	*Labyrinthula*	Atlantic Sheltered Bay (84)
Microspora (microsporans)	*Glugea stephani*	Continental Shelf with Flounder (58)
Myxozoa (myxozoans)	*Myxobolus cerebralis*	Salmon (56)
Oomycota (oomycotes; water molds)	*Saprolegnia parasitica*	Pond Bottom (102)
Paramyxea (paramyxeans)	*Marteilia refringens*	Estuary with Oyster Bar (54)
Phaeophyta (phaeophytes; brown algae)	*Fucus vesiculosus* *Nereocystis* (bladder kelp)	North Atlantic Coast (62) Cover illustration (iv)
Plasmodial Slime Molds	*Fuligo septica*	Birch Forest Floor (60)
Plasmodiophoromycota (plasmodiophorids)	*Plasmodiophora brassicae*	Cabbage Field (86)
Prymnesiophyta (prymnesiophytes)	*Emiliania huxleyi* *Prymnesium parvum*	Marine Chalk Cliffs (76) Marine Chalk Cliffs (76)
Raphidophyta (raphidophytes)	*Chattonella* *Vacuolaria*	River Delta (70) River Delta (70)
Rhizopoda (rhizopods)	*Amoeba proteus* *Arcella polypora* *Mayorella penardi*	Swamp (52) Swamp (52) Swamp (52)
Rhodophyta (rhodophytes; red algae)	*Polysiphonia harveyi*	North Atlantic Coast (62)
Xanthophyta (xanthophytes; yellow-green algae)	*Botrydiopsis* *Ophiocytium*	Lake Surface (78) Lake Surface (78)

PHYLUM (COMMON NAME)	REPRESENTATIVE(S)	HABITAT (PAGE NO.)

Kingdom Protoctista (continued)

Xenophyophora (xenophyophores)

Galatheammina tetradea — Marine Abyss (66)
Psammetta globosa — Marine Abyss (66)
Reticulammina lamellata — Marine Abyss (66)
Stannophyllum zonarium — Marine Abyss (66)

Zoomastigina (zoomastiginids)

Lophomonas (parabasalian) — Fallen Log (72)
Trichomonas (parabasalian) — Fallen Log (72)
Trichonympha (parabasalian) — Fallen Log (72)

Kingdom Fungi

Ascomycota (ascomycotes)

Morchella esculenta (morel) — Orchard (110)

Basidiomycota (basidiomycotes)

Amanita — Forest Clearing (112)
Amanita muscaria (fly agaric) — Cover illustration (iv)
Armillaria (honey mushroom) — Forest Clearing (112)
Boletus (edible bolete) — Forest Clearing (112)
Fomes (rusty-hoof fomes) — Forest Clearing (112)

Deuteromycota (deuteromycotes)

Titaeosporina — Forest Clearing (112)

Mycophycophyta (lichens)

Parmelia conspersa (boulder lichen) — Orchard (110)

Zygomycota (zygomycotes; mating molds)

Dactylaria — Orchard (110)
Dactylella tylopaga — Orchard (110)

Kingdom Animalia

Acanthocephala (acanthocephalans; spiny-headed worms))

Plagiorhynchus cylindraceus — Forest Soil Puddle (136)

Annelida (annelids)

Lumbricus terrestris (earthworm) — Forest Soil Puddle (136)

PHYLUM (COMMON NAME)	REPRESENTATIVE(S)	HABITAT (PAGE NO.)

Kingdom Animalia (continued)

Arthropoda (arthropods)	*Anopheles* (mosquito)	Mosquito and Blood (98)
	Armadillium vulgare (pill bug)	Forest Soil Puddle (136)
	Balanus (barnacle)	California Rocky Intertidal Zone (142)
	Parathemisto (amphipod)	Pacific Nearshore Waters (104)
	Plusiotis aurigans (golden beetle)	Costa Rican Tropical Forest at Night (144)
	Pterotermes occidentis (termite)	Fallen Log (72)
	Rothschildia lebeau (saturniid moth)	Costa Rican Tropical Forest at Night (144)
Brachiopoda (brachiopods)	*Terebratulina septentrionalis* (marine lamp shell)	Grassy Shoal (132)
Chaetognatha (chaetognaths; arrow worms)	*Sagitta bipunctata*	Open Ocean Rock and Sand Bottom (120)
Chordata (chordates)	*Bos taurus* (cattle)	Pasture (22)
	Enhydra lutris (sea otter)	California Rocky Intertidal Zone (142)
	Holocentrus xantherythrus (squirrel fish)	Coral and Squirrel Fish (130)
	Larus crassirostris (black-tailed gull)	Philippine Coral Sands (126)
	Lissoclinum (tunicate)	South Pacific Coral Reef (36)
	Manis tricuspis (African tree pangolin)	Cover illustration (iv)
	Onchorhynchus nerka (sockeye salmon)	Salmon (56)
	Perca flavescens (yellow perch)	Pond Bottom (102)
	Platiclothus stellatus (starry flounder)	Continental Shelf with Flounder (58)
	Sebastes serriceps (rock fish)	Shallow Pacific California Coast (138)
Cnidaria (cnidarians; coelenterates)	*Eunicella verrucosa* (coral)	Coral and Squirrel Fish (130)
	Hydra viridis	Lake Shore (40)
	Obelia	Caribbean Reef Seafloor (122)

PHYLUM (COMMON NAME)	REPRESENTATIVE(S)	HABITAT (PAGE NO.)

Kingdom Animalia (continued)

PHYLUM (COMMON NAME)	REPRESENTATIVE(S)	HABITAT (PAGE NO.)
Ctenophora (ctenophores; comb jellies)	*Beroe cucumi* (sea cucumber)	Open Ocean (124)
	Bolinopsis infundibulum (comb jelly)	Open Ocean (124)
	Cestum veneris (Venus's girdle)	Open Ocean (124)
	Pleurobrachia pileus (sea gooseberry)	Open Ocean (124)
Echinodermata (echinoderms)	*Ophiopholis aculeata* (daisy serpent star)	Sandy Seashore (128)
Echiura (echiurans; spoon worms)	*Listriolobus pellodes*	Sandy Marine Coast (140)
Ectoprocta (ectoprocts)	*Bugula*	Shallow Pacific California Coast (138)
Entoprocta (entoprocts)	*Barentsia*	Shallow Pacific California Coast (138)
Gastrotricha (gastrotrichs)	*Lepidodermella*	Grassy Shoal (132)
Gnathostomulida (gnathostomulids; jaw worms)	*Problognathia minima*	Grassy Shoal (132)
Hemichordata (hemichordates)	*Balanoglossus*	Sandy Marine Coast (140)
	Ptychodera flava	Open Ocean Rock and Sand Bottom (120)
	Saccoglossus kowalevskii	Sandy Marine Coast (140)
Kinorhyncha (kinorhynchs)	*Echinoderes kozloffi*	Pebbled Sea Bottom (118)
Loricifera (loriciferans)	*Nanaloricus mysticus*	Pebbled Sea Bottom (118)
	Pliciloricus enigmaticus	Pebbled Sea Bottom (118)
Mesozoa (mesozoans)	*Dicyema truncatum*	Philippine Coral Sands (126)

PHYLUM (COMMON NAME)	REPRESENTATIVE(S)	HABITAT (PAGE NO.)

Kingdom Animalia (continued)

PHYLUM (COMMON NAME)	REPRESENTATIVE(S)	HABITAT (PAGE NO.)
Mollusca (molluscs)	*Crassostrea virginica* (American oyster)	Estuary with Oyster Bar (54)
	Flabellinopsis iodinea (purple nudibranch)	California Rocky Intertidal Zone (142)
	Haliotus cracherodii (black abalone)	California Rocky Intertidal Zone (142)
	Loligo pealei (common squid)	California Rocky Intertidal Zone (142)
	Mytilus californianus (California mussel)	California Rocky Intertidal Zone (142)
	Nautilus	Philippine Coral Sands (126)
	Nautilus pompilius (chambered nautilus)	Cover illustration (iv)
	Octopus dofleini (Pacific giant octopus)	California Rocky Intertidal Zone (142)
	Stenoplax (chiton)	California Rocky Intertidal Zone (142)
Nematoda (nematodes)	free-living nematode	Lichen Water Film (134)
Nematomorpha (nematomorphs; Gordian worms; horsehair worms)	*Gordius villoti*	Forest Soil Puddle (136)
Nemertina (nemertines; ribbon worms)	ribbon worm	Coral and Squirrel Fish (130)
Onychophora (onychophorans; velvet worms)	*Peripatus*	Cover illustration (iv) Costa Rican Tropical Forest at Night (144)
Pentastoma (pentastomes)	*Armillifera*	Philippine Coral Sands (126)
Phoronida (phoronids)	*Phoronopsis vancouverensis*	Shallow Pacific California Coast (138)
Placozoa (placozoans)	*Trichoplax*	Pebbled Sea Bottom (118)
Platyhelminthes (platyhelminths; flatworms)	flatworm	Sandy Seashore (128)

PHYLUM (COMMON NAME)	REPRESENTATIVE(S)	HABITAT (PAGE NO.)

Kingdom Animalia (continued)

PHYLUM (COMMON NAME)	REPRESENTATIVE(S)	HABITAT (PAGE NO.)
Porifera (poriferans; sponges)	*Euplectella speciosissima*	Open Ocean Rock and Sand Bottom (120)
Priapulida (priapulids)	*Meiopriapulus*	Philippine Coral Sands (126)
Rotifera (rotifers)	*Philodina*	Lichen Water Film (134)
Sipuncula (sipunculans; peanut worms)	*Themiste lageniformis*	Sandy Marine Coast (140)
Tardigrada (tardigrades; water bears)	*Echiniscus arctomys*	Lichen Water Film (134)
Vestimentifera (vestimentiferans; tube worms)	*Riftia pachyptila*	Marine Abyss (146)

N = 33

Kingdom Plantae

PHYLUM (COMMON NAME)	REPRESENTATIVE(S)	HABITAT (PAGE NO.)
Angiospermophyta (angiosperms)	*Alnus* (alder)	Woodland Stream (46)
	Brassica oleracea (cabbage)	Cabbage Field (86)
	bromeliad	Rain Forest Canopy (162)
	Cercidium (paloverde)	Fallen Log (72)
	orchid	Rain Forest Canopy (162)
	Ulmus americana (American elm)	Cover illustration (iv)
	Utricularia (bladderwort)	Pond Rocks (94)
	Zostera marina (eelgrass)	Atlantic Sheltered Bay (84)
		Grassy Shoal (132)
Bryophyta (bryophytes)	*Polytrichum juniperinum*	Woods (152)
Coniferophyta (conifers)	*Pinus rigida* (pitch pine)	Pitch Pine Barrens (156)
Cycadophyta (cycads)	*Zamia* (coontie)	Florida Bush (154)
Filicinophyta (ferns)	*Osmunda cinnamomea* (cinnamon fern)	Woods (152)

PHYLUM (COMMON NAME)	REPRESENTATIVE(S)	HABITAT (PAGE NO.)

Kingdom Plantae (continued)

PHYLUM (COMMON NAME)	REPRESENTATIVE(S)	HABITAT (PAGE NO.)
Ginkgophyta (ginkgo)	*Ginkgo biloba*	Hillside in China (158)
Gnetophyta (gnetophytes)	*Welwitschia mirabilis*	African Desert (160)
Lycopodophyta (lycopods; club mosses; ground pines)	*Lycopodium obscurum* (tree club moss)	Woods (152)
Psilophyta (psilophytes)	*Psilotum* (whisk fern)	Florida Bush (154)
Sphenophyta (sphenophytes)	*Equisetum arvense* (common horsetail)	Pitch Pine Barrens (156)

Glossary

Abbreviations used in the Glossary:

abbr.	abbreviation	s.	singular
adj.	adjective	usu.	usually
esp.	especially	v.	verb
pl.	plural	var.	variation

aboral (adj.) Away from the mouth.

abyss Deep ocean; the organisms and material usu. found beyond the continental shelf.

acervulus (pl. acervuli) Mat of hyphae that gives rise to conidiophores packed together closely to form a bedlike mass.

adult Fully developed and mature individual capable of producing sex cells (eggs, sperm, or pollen) that can fuse to form an embryo.

aerobe Organism that lives in the presence of and uses oxygen. Obligate aerobes are unable to live without oxygen; facultative aerobes can live in oxic or anoxic environments.

agamont Adult life-cycle stage that is capable of reproduction but does not produce gametes.

agar Hardening substance for cultivating bacterial, protoctist, and fungal microorganisms; constituent of some gels used for electrophoresis that is prepared from a gelatinous substance (agar-agar) extracted from red algae.

agglutination Formation of clumps of cells, esp. pollen, bacteria, red blood cells, spermatozoans, and some protoctists, either spontaneously or after treatment with a specific antibody or other agent.

aggregate (v.) To form a cluster (e.g., of organisms or cells).

algae (s. alga) Heterogeneous group of eukaryotic, unicellular, colonial, or multicellular aquatic organisms that are photosynthetic at some stage in their life history. (Photosynthesis occurs in plastids.) Although traditionally classified as plants, the major groups of algae are now phyla in the kingdom Protoctista.

alternation of generations Reproductive cycle in which haploid phases alternate with diploid phases.

amastigote Microorganism or life-cycle stage of an organism that lacks undulipodia.

ameba (var. amoeba, pl. amoebae) (1) Unicellular, protoctist life-cycle stage that moves by means of pseudopods and whose shape is therefore subject to constant change. (2) Informal name for a member of the protoctist phylum Rhizopoda.

ameboid (adj.; var. amoeboid) Resembling an ameba in shape, properties, or mode of movement.

amebomastigote (1) Ameba that undergoes a transformation to a mastigote stage. (2) Informal name of a member of the zoomastiginid class Amebomastigota.

amino acid An organic acid containing the amino group (NH_2) and a carboxyl group (–COOH); subunit of proteins.

anaerobe Organism living in the absence of oxygen. Obligate anaerobes are unable to live in even low concentrations of oxygen; facultative anaerobes live in oxic or anoxic environments; aerotolerant organisms can live in the presence of oxygen but do not use it.

analogous (adj.) Convergent; relating to structures or behaviors with the same function that have not evolved from common ancestors (e.g., the wings of insects and bats). Compare *homologous*.

anastomosis Formation of a network by the fusion of branches, filaments, or tubes.

angiosperm Plant that produces its seeds inside flowers; commonly called a flowering plant; member of the plant phylum Angiospermophyta.

anoxic (adj.) Devoid of molecular oxygen (referring to habitats).

antenna (pl. antennae or antennas) Movable, usu. segmented sensory organ on the head of an organism.

antennule Small antenna or similar appendage.

antheridium (pl. antheridia) Multicellular male sex organ; sperm-producing gametangium of plants other than seed plants.

anthrax Disease affecting the lungs of warm-blooded animals that is caused by the bacterium *Bacillus anthracis*.

antibiotic Substance produced by organisms (typically fungi or bacteria) that injures, kills, or prevents the growth of other organisms (typically bacteria).

antibody Protein produced by vertebrate blood cells that is capable of defending the animal against a specific virus, bacterium, biochemical, or other imposition.

anus Posterior opening of the animal digestive tract, through which fecal waste is excreted.

apical complex Specialized structure at the apex of an apicomplexan that facilitates attachment and penetration of the organism into the tissue cell of the animal in which the apicomplexan resides.

archaebacteria (s. archaebacterium) Class of prokaryotes that includes thermoacidophiles, halophiles, and methanogens, typically found in extreme environments (e.g., hot springs or salt lakes). Some scientists think that archaebacteria are the most ancient group of organisms still living.

archegonium (pl. archegonia) Multicellular female sex organ; egg-producing gametangium of plants other than seed plants.

ascospore Spore contained in an ascus formed by karyogamy followed by meiosis.

ascus (pl. asci) A saclike cell of a hypha that contains a definite number of ascospores (usu. eight).

astropyle Main opening of the central capsule of phaeodarian actinopods.

ATP Adenosine triphosphate; molecule that is the primary energy carrier for cell metabolism and motility.

autogamy Self-fertilization; the union of two nuclei, both of which are derived from the nucleus of a single parent.

autopoiesis Organismal self-maintenance.

autotrophy Mode of nutrition in which organic compounds are grown and synthesized from inorganic compounds by organisms that use energy from sunlight or from the oxidation of inorganic compounds.

axoneme Microtubule or shaft of microtubules extending the length of an undulipodium, pseudopod, or axopod.

axopod Permanent pseudopod stiffened by a microtubular axoneme.

bacillus Bacterium that is rod-shaped.

basal cell Cell in the lowest layer of stratified tissue (such as epidermis and other epithelia) from which the tissue is renewed.

basal disc Platelike structure at the base of a cell process; part of the sporophore of a dictyostelid slime mold.

basidiocarp Mushroom or other reproductive structure of a fungus that bears basidia.

basidiospore Spore borne on a basidium that results from karyogamy and meiosis.

basidium (pl. basidia) Club-shaped structure bearing basidiospores on its surface.

benthic (adj.) Of, relating to, or occurring at the bottom of a body of water.

binary fission Reproduction in which one parent cell divides into two offspring cells of roughly equal size.

binomial nomenclature System of naming and identifying organisms that assigns each organism two names, a genus name and a species name.

binucleate (adj.) Having two nuclei.

bioassay Determination of the relative strength of a substance by comparing its effect on a test organism with the effect of a standard preparation.

biochemical Substance produced by chemical reaction in a living organism.

biodiversity The total number of kinds of and the differences between living species.

biogenic (adj.) Produced by living organisms or their remains.

bioluminescence Emission of biochemically generated light by living organisms.

biome Huge, contiguous territory that is geographically definable (e.g., the world ocean, the northern tundra).

biomineralization Formation of minerals by living organisms.

biosphere The biota and all of its habitats; the volume at the surface of Earth that harbors life.

biota All living matter on Earth at a given time; the flora, fauna, and microbiota taken together.

blastopore Opening connecting the cavity of the gastrula stage of an embryo with the outside; the future mouth of some animals and the anus of others.

blastula Animal embryo; small, hollow ball of cells that usu. goes on to develop into a gastrula, an embryo with the beginnings of tissues.

bothrosome Organelle on the membrane of labyrinthulomycotes that produces new membrane, sequesters calcium, and filters cytoplasm for the production of the proteinaceous, extracellular slime-net matrix. Sagenogen.

brackish (adj.) Of or relating to water with a salinity between that of sea water and that of fresh water.

bract Modified, often colored leaf beneath a flower or flower cluster.

broadleaf (adj.) Having broad leaves (as opposed to needles).

Brownian motion Random movement of tiny particles in solution, such as the components of cells.

bryophyte Member of a group of nonvascular plants that consists of the mosses, hornworts, and liverworts.

budding Propagule formation; asexual reproduction by the outgrowth of a protrusion (bud) from a parent cell or body.

Burgess Shale Mid-Cambrian sedimentary rock unit of British Columbia that contains extraordinarily well preserved animal fossils.

calcareous (adj.) Containing calcium, usu. in the form of calcium carbonate ($CaCO_3$).

calyx (pl. calyxes or calyces) (1) In animals, cup-shaped structure of crinoids and entoprocts. (2) In plants, the sepals; the cup-shaped outer series of two series of floral leaves.

Cambrian period Earliest geological period of the Paleozoic era, from about 590 million to 505 million years ago, during which many phyla of multicellular animals first arose.

capsular wall Wall of a spherical or nearly spherical structure.

carbohydrate Compound composed of carbon, oxygen, and hydrogen of the general formula $C_x(H_2O)_y$. Carbohydrates include sugars (monosaccharides and disaccharides) and their derivatives and polysaccharides, such as starch and cellulose.

Carboniferous period Geological period of the late Paleozoic era, from about 350 million to 285 million years ago, during which coal beds were formed.

carboxysome Organelle inside plastids that is thought to contain the CO_2-fixing enzyme ribulose bisphosphate carboxylase.

carnivory Mode of nutrition by which an organism obtains nutrition and energy by eating live animals.

carotenoid Red, orange, or yellow isoprenoid pigment (e.g., carotene, xanthophyll) found in plastids.

cartilage Translucent, elastic, skeletal connective tissue.

cell Basic structural building block of living organisms, consisting of protoplasm bounded by a membrane and—in plants, bacteria, and fungi—also surrounded by a nonliving rigid wall.

cell process See *process*.

cellulose Polysaccharide composed of glucose units; chief constituent of the cell wall in plants and chlorophytes.

centimeter (abbr. cm) 0.01, or 10^{-2}, meter. (2.56 cm = 1 inch.)

centric (adj.) Of or relating to radial symmetry of the valves in diatoms.

centriole Eukaryotic cell organelle; barrel-shaped organelle 0.25 micrometer in diameter that is composed of a [9(3)+0] array of microtubules. Centrioles appear during animal and some protoctist cell divisions at each pole of the mitotic spindle. Centrioles are absent at the poles of mitotically dividing cells of fungi, plants, and many protoctists.

centriole-kinetosome [9(3)+0] microtubular structure that is naked (centriole) or is at the base of an undulipodium (kinetosome).

centromere Kinetochore; structure on each chromosome that attaches it to microtubules of the mitotic spindle.

chaeta (pl. chaetae) Bristle; seta.

chasmolithic (adj.) Living inside fissures and cracks in rocks; referring to the colonization of surfaces of rock by microorganisms incapable of dissolving the rock.

chemotrophy Mode of nutrition in which energy is obtained—either from inorganic sources (chemoautotrophy) or from organic sources (chemoheterotrophy)—by chemical reactions independent of light.

chitin Tough, resistant, nitrogen-rich polysaccharide that is a component of exoskeletons (as in arthropods) and cell walls (as in some protoctists and fungi).

chlorophyll Green pigment that absorbs visible light energy and helps convert it to usable chemical energy in photosynthesis.

chloroplast Green plastid; plastid that contains chlorophylls *a* and *b* and is the site of photosynthesis.

cholesterol Steroid alcohol with the chemical formula $C_{27}H_{45}OH$ that is a major component of membranes of eukaryotic organisms.

chromatin Complex of nucleic acid (DNA) and basic proteins (histones) of which chromosomes are made during mitotic cell division.

chromosome Organelle inside the nucleus of eukaryotic cells that is made of chromatin and that contains most of the DNA (genetic material).

chrysoplast Yellow plastid; membrane-bounded photosynthetic organelle of chrysophytes, diatoms, and prymnesiophytes that contains chlorophylls *a* and *c*.

chytrid body Structure of chytridiomycotes in which zoospores form.

cilium (pl. cilia) Short undulipodium; intracellular but protruding organelle of motility composed of microtubules in the [9(2)+2] configuration and underlain by the [9(3)+0] kinetosome from which it develops.

class Taxonomic level below phylum and above order.

clast Rock particle or fragment.

cloaca Exit chamber common to the digestive system (the gastrointestinal tract), the reproductive system, and the urinary system from which feces, gametes, and liquid waste are excreted.

coccoid Spherical, nearly spherical, or berry-shaped structure.

coccolith External, platelike structure on some prymnesiophytes that is made of calcium carbonate.

coccolithophorid Prymnesiophyte that bears coccoliths.

coelenteron (pl. coelentera) Hollow digestive cavity that is characteristic of cnidarians.

coelom Body cavity that encloses the vascular system and digestive tract; characteristic of nearly all animal phyla (except acoelomates and pseudocoelomates).

coenocytic (adj.) Having more than one nucleus in common cytoplasm (referring to cells or organisms). Multinucleate; syncytial; plasmodial.

collar cell Cell with a single undulipodium that generates currents by which poriferans draw water through their ostia and catch food particles.

colonial (adj.) Relating to a group of cells or organisms of the same species. Although each is capable of growth by division, colony members live in stable but loose association.

commensalism Physical, nonnecrotrophic association between members of two or more species in which neither species necessarily takes nutrients from the other.

community Set of populations of organisms of different species in the same place at the same time.

cone See *strobilus*.

conidiophore Specialized hypha that bears conidia.

conidium (pl. conidia) Propagule of fungi; mitotically produced spore borne on a conidiophore or on a nonspecialized hypha that is capable of further growth in the absence of sex and fusion.

conifer Cone-bearing plant of the phylum Coniferophyta.

conjugation In prokaryotes, cell-to-cell contact between a donor and a recipient bacterium at which the transmission of genetic material occurs. In eukaryotes, fusion of nonundulipodiated gametes or gamete nuclei.

cortex Outer layer of an organism or organ.

cotyledon Leaflike structure of plant seeds that provides food for the developing embryo.

crinoid Member of a class of the phylum Echinodermata that includes sea lilies and feather stars.

cristae (s. crista) Tubular or pouchlike and inwardly-directed folds of the inner membrane of a mitochondrion that are the site of ATP production during aerobic metabolism.

crypt Pit or depression (gullet) characteristic of cryptomonads.

cryptobiosis Dormant state of a propagule; suspended or deathlike condition usu. induced by starvation, desiccation, or extreme temperatures.

crystalline style Enzyme-releasing organ of the digestive system of bivalve molluscs.

cuticle Outer layer or covering, usu. of a plant or an animal, that is composed of metabolic products rather than of cells.

cyanelle Intracellular structure, considered to be a cyanobacterial symbiont or an organelle derived from symbiotic cyanobacteria, that, containing thylakoids, is active in oxygenic photosynthesis.

cyclosis See *cytoplasmic streaming.*

cyst Propagule; encapsulated form, often a dormant stage, of one of several types of organisms that forms in response to extreme environmental conditions.

cytology Study of cells.

cytoplasm Fluid portion of a cell, exterior to the nucleus or nucleoid and containing enzymes and metabolites in solution.

cytoplasmic streaming Circulation of cytoplasm in the cell; characteristic intracellular motility of eukaryotic cells. Cyclosis; protoplasmic streaming.

cytosome Cell mouth; ingestive opening of protists (e.g., euglenids).

deciduous (adj.) Shed or sloughed off seasonally or at a certain stage in the life cycle; having deciduous parts (e.g., the leaves of temperate-zone trees).

dehiscence Opening of a structure by drying or by programmed death of certain structures or cells (e.g., to allow propagule escape).

deposit feeding Act or process of eating material that has settled or has been deposited on the bottom of a body of water.

desmosome Intercellular membranous junction fastening cells together in animal tissues.

detritus Loose natural material, such as rock fragments or organic particles, that results directly from disintegration of rocks or organisms; organic-rich clastic sediment.

dicotyledon (var. dicot) Plant that has two cotyledons in the seed; member of a subphylum of angiosperms.

differentiation Process characteristic of animal and plant development in which unspecialized cells become tissue components specialized for different functions.

dikaryon Cell or hypha that contains any number of nuclei of two genetically distinct forms (the sources of which may or may not be known), usu. one from each parent after sexual fusion; typical of fungi. Compare *heterokaryon*.

dikaryotic (adj.) Containing two nuclei, each usu. from a different parent.

dioecious (adj.) Having male and female organs on different individuals of the same species.

diploid (adj.) Having two complete sets of chromosomes (the chromosome complement of sexually produced eukaryotic cells), one each from the maternal and paternal parents. Compare *haploid*.

DNA Deoxyribonucleic acid; a long molecule composed of nucleotides in a linear order that constitutes the genetic information of cells and that is capable of replicating itself and of synthesizing RNA.

dormant (adj.) Nongrowing; resistant; in a state of suspended activity (referring to propagules, tissues, or organisms).

dorsal (adj.) Toward the back side.

ecology Study of relationships between organisms and their environment.

ecosystem Community of organisms in their environment such that the rate of flow of carbon, nitrogen, sulfur, phosphorus, and other biologically important elements is greater inside of the system than it is between systems.

ectomycorrhiza (pl. ectomycorrhizae) Symbiosis between plant roots and soil fungi, typical of basidiomycotes and trees, in which the fungi cover the plant root with a mantle but do not penetrate the root cell.

ectoplasm Outermost, relatively rigid, transparent layer of cytoplasm.

Ediacaran biota Protoctist and animal fossils from Ediacara, South Australia, and some two dozen other fossil localities worldwide that are about 700 million years old.

egg Female gamete, which is nonmotile and usu. larger than the male gamete.

elater Hygroscopic cell or band, usu. attached to the spore (e.g., of a horsetail or moss), that aids in dispersing spores

electrophoresis Gel electrophoresis; laboratory technique used to separate macromolecules on the basis of electrical charge or size.

embryo Early developmental stage of a multicellular organism (animal or plant) that develops from a zygote (fertilized egg).

encyst (v.) To form or become enclosed in a cyst.

endobiotic (adj.) Endosymbiotic; of or relating to an association in which one partner lives within the other partner.

endolithic (adj.) Living inside of rocks, in limestone, by active penetration of the rock by microorganisms capable of dissolving the rock.

endomycorrhiza (pl. endomycorrhizae) Symbiosis between plant roots and soil fungi, typical of zygomycotes and herbaceous plants, in which the fungi do not cover the plant roots with a mantle but do penetrate the root cells.

endoplasm Inner, relatively fluid central portion of cytoplasm of eukaryotic cells.

endoplasmic reticulum ER; extensive endomembrane system found in most protoctist, animal, and plant cells.

endosperm Tissue surrounding a plant embryo in a seed that contains stored food.

endospore Desiccation- and heat-resistant spore produced inside bacteria.

endosymbiosis Association of partners, members of different species, in which one organism lives inside the body or cell of another.

enzyme Protein catalyst for a specific substrate and product; biochemical that accelerates but does not enter metabolic reactions.

epibiosis Association of organisms in which one lives on the surface of another.

epidermis Outer layer of skin or of leaves.

epiphyte Plant that lives supported, but not nourished, by a plant of another species; epibiotic plant.

epithelium (pl. epithelia) Tissue that covers the inner or outer surface of a body or structure.

esophagus (pl. esophagi) Muscular tube through which food is passed from the pharynx to the stomach; gullet.

etioplast Undeveloped plastid that lacks chlorophyll; typical of plants grown in the dark.

eubacteria (s. eubacterium) All bacteria other than the archaebacteria. Eubacteria have distinctive lipids, RNAs, and RNA polymerases, and most have cell walls that contain muramic acid.

eukaryote Organism composed of cells with membrane-bounded nuclei, chromatin organized into more than one chromosome, and usu. organelles (such as mitochondria and plastids).

excyst (v.) To exit the cyst stage.

exoskeleton External supportive covering.

eyespot Small, pigmented, and probably light-sensitive structure in certain undulipodiated protists.

facultative (adj.) Optional; exhibiting a certain life-style under some environmental conditions but not others.

family Taxonomic level below order and above genus.

fat Compound composed of carbon, hydrogen, and oxygen with a lower proportion of oxygen to carbon than that of carbohydrates; the major form of lipids in some animals and in some plants.

fauna Animal life. (Inappropriate term for protoctists and bacteria.)

feces Solid waste from the gastrointestinal tract of animals.

fermentation Anaerobic respiration; mode of nutrition in which organic compounds are degraded in the absence of oxygen, with organic compounds serving as terminal electron acceptors, yielding energy and organic end products.

fertilization Fusion of two haploid cells, gametes, or gamete nuclei to form a diploid nucleus, diploid cell, or zygote.

fetal (adj.) Relating to a fetus, the unborn or unhatched vertebrate after it has attained its basic structural plan.

fibril Thread-shaped solid structure; filament.

flagellin Polymeric protein that is the main constituent of flagella.

flagellum (pl. flagella) (1) In prokaryotes, the long, thin, solid, extracellular structure of motility composed of one of a number of flagellin proteins. (2) In eukaryotes, a confusing term for an undulipodium; the long, fine, intrinsically motile intracellular structure used for sensing, for locomotion, for feeding, and so on, that is underlain by microtubules composed of tubulin and other nonflagellin proteins.

flora Plant life. (Inappropriate term for protoctists or fungi.)

food web Nutritional hierarchy in a natural community such that each organism in the web feeds on an organism below it and is eaten by an organism above it, ranging from the simplest autotroph at the bottom to the carnivore at the top.

fossil The bodily remains or traces of an organism from past geological ages that have been preserved in Earth's crust.

free-living (adj.) Living unattached, or attached but not symbiotrophically, to other organisms.

frond (1) Large leaflike structure, usu. with many divisions (as in ferns or palms). (2) Divided thallus that resembles a leaf (as in seaweed).

fruit Mature, ripened ovary (or ovaries), the seed-bearing structure, and associated structures of an angiosperm.

fruiting body Structure that contains or bears seeds, spores, cysts, or other propagules. (Ill-advised botanical term.)

fucoxanthin Brown carotenoid found in chrysophytes, phaeophytes, and other algae.

Gaia The whole Earth living system; the biota—the sum of the microbiota, plants, and animals—embedded in the biosphere. The Vernadskian space (see page 18) where life exists.

gametangium (pl. gametangia) Organ of plants or protoctists in which mitosis occurs and gametes or gamete nuclei are generated, and from which they are released.

gamete Mature haploid reproductive cell capable of fusion with another gamete, of a different mating type, to form a diploid zygote.

gametophyte Haploid gametangium and gamete-producing generation in plants and some protoctists.

gamont Body of an organism; adult life-cycle stage that produces gametes.

gap junction Intercellular, membranous, discontinuous junction that fastens cells together in animal tissues and that is thought to regulate the flow of ions between cells.

gastrodermal (adj.) Relating to the thin layer of tissue that lines the digestive tract of marine animals.

gastrula (pl. gastrulae or gastrulas) Animal embryo in which the blastula with its single layer of cells becomes a three-layered embryo and the inner digestive tissue begins to form (a process called gastrulation).

gemma (pl. gemmae; var. gemmule) Propagule; reproductive structure; a small mass of cells from body tissue that can be released and develop into a new individual.

gemmule See *gemma.*

gender The sum of the behavioral and physiological traits that identify potential mating types (e.g., sexual partners or conjugants) prior to a sexual union.

gene Smallest physical unit of heredity; sequence of nucleotides in DNA that is sufficient to specify a protein product.

generative (adj.) Capable of further growth and/or reproduction.

genome Complete set of genetic material required for life of a cell or organism.

genus (pl. genera) Taxonomic level below family and above species; the first part of the two-part name in binomial nomenclature.

germ tube Tube-shaped structure capable of continued growth.

germinate (v.) To begin to grow.

giant cell Unusually large cell among cells of normal size.

gill (1) One of the plates on the underside of the cap of a basidiomycote. (2) Respiratory organ used for uptake of oxygen and release of CO_2 by aquatic animals.

glucose Carbohydrate with the chemical formula $C_6H_{12}O_6$.

Golgi apparatus Layered, cup-shaped organelle composed of modified endoplasmic reticulum that plays a role in storing and secreting metabolic products. Golgi body; dictyosome (referring to plants).

Golgi body See *Golgi apparatus.*

gonad Animal organ composed of tissues that produce gametes; e.g., the female ovary (which produces eggs) and the male testis (which produces sperm).

gonomere Reproductive structure of ellobiopsids, borne on a trophomere, that contains spores.

Gram-negative (adj.) Failing to retain the purple stain when subjected to the Gram staining method, indicating the presence of certain component layers in the cell wall (e.g., Gram-negative bacteria). The result of Gram staining is a characteristic used to classify bacteria.

granellare The plasma body (protoplasm) of a xenophyophoran together with its surrounding branched, yellowish tubes.

grex Multicellular, migratory slug phase of dictyostelid slime molds.

gullet Oral cavity.

gymnosperm Seed plant in which the seeds are not enclosed in an ovary (e.g., a conifer).

habitat Immediate surroundings of a population or community.

haploid (adj.) Having one set of chromosomes, the chromosome complement of sexually produced eukaryotic cells after meiosis. Compare *diploid*.

haplosporosome Spherical, membrane-bounded organelle, of haplosporidians and possibly myxozoans and paramyxeans, with unknown function.

haptoneme Intracellular, protruding, microtubular organelle of prymnesiophytes, usu. coiled, often used as a holdfast.

herbaceous (adj.) Nonwoody (referring to plants).

herbivory Mode of nutrition by which an organism obtains nutrition and energy by eating plants or algae.

hermaphrodite Organism that simultaneously possesses male and female organs on the same body.

heterokaryon Cell or hypha that contains any number of nuclei of two genetically distinct forms (the sources of which are known), usu. one from each parent after sexual fusion; typical of fungi. Compare *dikaryon*.

heterokont (adj.) Having two undulipodia of unequal length, usu. one forward-directed and the other trailing (referring to mastigote or algal cells).

heterosporous (adj.) Of or relating to a diploid plant body capable of forming haploid spores as products of meiosis, of two kinds (typically, microspores and megaspores) simultaneously.

heterotrophy Mode of nutrition in which carbon and energy are obtained from organic compounds from, for example, the bodies of other organisms, detritus, or solutions of organic compounds that have been consumed osmotrophically. Ultimately the organic compounds are produced by autotrophs.

Higgin's larva Juvenile stage of loriciferans.

histone Member of a class of proteins rich in nitrogenous bases that complexes with nuclear DNA in eukaryotes.

homologous (adj.) Relating to structures or behaviors that have evolved from common ancestors, even if the structures or behaviors have diverged in form and function.

homosporous (adj.) Relating to a plant body that is capable of forming haploid spores as products of meiosis, of only one kind.

hormone Substance that is secreted directly into body fluid or blood by ductless glands and is carried to a specific target cell or organ, where, in minute amounts, the hormone brings about a specific physiological response.

host Nonspecific term for an organism that provides nutrition or lodging for symbiotrophs, or the larger member of a symbiotrophic association. (Should be replaced by a more precise term.)

hydrostatic (adj.) Relating to fluids at rest or to the pressures they exert or transmit.

hydrothermal (adj.) Relating to hot water.

hygroscopic (adj.) Readily taking up and retaining moisture.

hypertrophy Overgrowth; unusual increase in size or number of cells or organisms.

hypha (pl. hyphae) Threadlike tubular filament, a component of a mycelium, usu. of a fungus or a funguslike protoctist.

hypnocyst Resting cyst of a dinomastigote. A hystrichosphere is a fossil hypnocyst.

infection Initiation of a symbiotrophic (usu. necrotrophic) relationship between organisms of different species.

infusion Liquid extract obtained by steeping or soaking a substance, usu. leaves or other dry organic material, in usu. hot or boiling water.

infusoriform larva Ciliated, immature mesozoan; a swarm larva.

inoculation (1) Introduction of bacteria, protists, or fungi into a medium suitable for their growth. (2) Introduction of material from one organism into a competent animal that stimulates the production of antibodies in the latter.

inorganic (adj.) Lacking carbon–hydrogen bonds.

intertidal (adj.) Relating to or occuring in the zone between the tides, which is covered with sea water at high tide and exposed at low tide.

introvert Slender anterior body part that can be turned inside out and completely withdrawn into the trunk of an animal body.

invertebrate Ill-advised term for all animals that are not members of the phylum Chordata, subphylum Vertebrata (e.g., animals lacking a spinal column).

ion Atom or group of atoms that carries a positive or negative electric charge.

isoprenoid One of a class of organic compounds that are synthesized from multiples of a ubiquitous five-carbon compound precursor (isopentenyl pyrophosphate).

jacket cells Ciliated cells that comprise the outer layer of cells in mesozoans.

karyogamy Fusion of nuclei, usu. in fertilization that precedes meiosis.

kinetid Kinetosome and associated microtubules and fibers in all undulipodiated cells (e.g., the unit of structure of the ciliate cortex).

kinetosomal fiber Fiber found proximally to cilia and undulipodia; part of the kinetid and therefore diagnostic for ciliate taxonomy.

kinetosome Organelle at the base of an undulipodium that is responsible for its formation and that is composed of microtubules in the [9(3)+0] configuration; centriole without an axoneme.

kinety Row of kinetids; structure of the ciliate cortex.

kingdom The most inclusive taxonomic level, immediately above phylum.

lamella (pl. lamellae) Flat, thin scale or flattened saclike structure.

larva (pl. larvae) Immature form of an animal, morphologically distinguishable from the adult form.

lasso cell Sticky, threadlike cell, found on the tentacles of cnidarians and ctenophores, that stings and captures prey.

legume Plant that is a member of the pea or bean family (Leguminosae).

leucoplast Cell organelle of algae; colorless or white plastid that often stores starch.

life cycle Sum of the events throughout the development of an individual organism that correlate environment and morphology with genetic and cytological observations (e.g., ploidy of the nuclei, fertilization, meiosis, karyokinesis, or cytokinesis).

life history Sum of the events throughout the development of an individual organism that correlate environment with changes in external morphology, formation of propagules, and other observable aspects (e.g., spore, ameba, pseudoplasmodium, slug, or sporophore).

lignin Polymer related to cellulose that is a major constituent of the secondary cell walls of most vascular plants.

lipid Member of a class of chemicals made by living beings, organic compounds soluble in organic but not aqueous solvents. Lipids include fats, waxes, steroids, phospholipids, carotenoids, and xanthophylls.

littoral zone Marginal zone of the sea or lake shore in which light reaches the bottom.

lobate (adj.) Having lobes.

lophophore Ridge surrounding the mouth that bears hollow, ciliated tentacles and traps food; found in several phyla of animals.

lorica Secreted protective covering, test, shell, valve, or sheath.

macrocyst Propagule; large cyst (e.g., of dictyostelids or myxomycotes).

macronucleus (pl. macronuclei) The larger of the two kinds of nuclei in ciliate cells; site of messenger RNA synthesis; "physiological" nucleus that contains many copies of each gene and, unlike the micronucleus, is required for growth and division.

madreporite External aboral terminus of the water vascular system of an echinoderm. Sieve plate.

mantle Covering or coat of brachiopods and molluscs; body wall that secretes a shell.

mastigoneme Hairlike, lateral projection on undulipodia.

mastigote Eukaryotic microorganism motile via undulipodia; eukaryotic "flagellated" cell.

mating Contact or fusion of cells, nuclei, or gamonts of complementary gender.

medulla (pl. medullae) Inner portion of a gland or other structure surrounded by cortex.

medusa Free-swimming, bell-shaped or umbrella-shaped stage in the life cycle of many cnidarians.

megagametophyte Female gametophyte—i.e., the haploid generation of the plant (e.g. in angiosperms)—located within the ovule of the seed.

megaphyll Usu. large leaf with several to many veins.

megasporangium (pl. megasporangia) Organ of plants in which female meiotic products form; usu. produces one to four megaspores.

megaspore Cell in the life cycle of plants; haploid spore that develops into a female gametophyte.

meiosis One or two successive divisions of a diploid nucleus (with two sets of chromosomes) that result in the production of haploid nuclei (with one set of chromosomes each).

merozoite Life-cycle stage (e.g., in apicomplexan protists) that is the mitotic product of a trophozoite.

mesokaryotic (adj.) Relating to dinomastigote nuclei that lack conventional histones and have permanently condensed chromosomes; literally, "between prokaryotic and eukaryotic."

Mesozoic era Geological time from 245 million to 66 million years ago that is composed of three periods: Triassic, Jurassic, and Cretaceous.

metabolism Sum of the enzyme-mediated biochemical reactions that occur continually in cells and organisms and provide the material basis for autopoiesis.

metabolite Small carbon compound that is the substrate, intermediate, or product of a metabolic reaction.

metamorphosis Discontinuous transformation from an immature to an intermediate or adult form (e.g., tadpole to frog).

metazoan Animal; organism classified in the kingdom Animalia.

meter (abbr. m) Basic metric unit of length; equal to 39.37 inches.

methanogenesis Production of methane gas (CH_4) by live organisms (bacteria).

methylotrophy Nutritional mode in some archaebacteria that involves the consumption of methane or other single-carbon compounds using oxygen as a source of carbon and energy.

microaerobic (adj.) Relating to environmental conditions in which oxygen is present in less than normal atmospheric concentrations (i.e., less than 20% by volume). Microoxic; dysaerobic.

microbe Organism that requires a microscope to be visualize it; microscopic living thing: bacterium, protist, or small fungus.

microbial mat Carpetlike community of microorganisms, usually cyanobacteria; living precursor of a stromatolite.

microbiota Microbial life; sum of the microorganisms in a given habitat.

microcosm The world of the subvisible; communities of organisms that can be seen only by microscopy.

microgametophyte Male gametophyte; the smaller haploid plant that produces male gametes by mitosis.

micrometer (abbr. μm) 10^{-6} meter.

micronucleus (pl. micronuclei) The smaller of the two kinds of nuclei in ciliate cells; does not synthesize messenger RNA; most are diploid and are required for meiosis and autogamy but not for asexual growth and division.

microorganism See *microbe*.

microphyll Small leaf with one vein; characteristic of lycopods.

microsporangium (pl. microsporangia) Structure on the male gametophyte in which male meiotic products form; produces many microspores.

microspore Haploid spore that develops into a male gametophyte.

microtubular ribbon Row of microtubules; component of kinetids, associated with the movement of ciliates.

microtubule Slender, hollow, proteinaceous, intracellular structure, usu. about 24 nm in diameter, found in axopods, axonemes, mitotic spindles, undulipodia, haptonemes, nerve cell processes, and other intracellular structures.

millimeter (abbr. mm) 0.001, or 10^{-3}, meter.

mitochondrion (pl. mitochondria) Organelle in which the chemical energy in reduced organic compounds (food molecules) is transferred to ATP molecules by oxygen-requiring respiration.

mitosis Nuclear division in which attached pairs of duplicate chromosomes move to the equatorial plane of the nucleus, separate at their centromeres, and move along the mitotic spindle to form two separate chromosome groups; subsequent division of the cell produces two genetically identical offspring cells.

mitotic spindle Transient microtubular structure that forms between the poles of nucleated cells, forming the structures along which chromosomes move in mitosis.

monad Single unit; single-celled organism.

monocotyledon (var. monocot) Plant that has only one cotyledon in the seed; member of a subphylum of angiosperms.

monoecious (adj.) Having male and female sex organs on the same individual. Hermaphroditic.

monomer Unit of a polymer; chemical component (e.g., amino acid of protein).

monosaccharide A simple sugar.

morphology Form and structure of an organism or any of its parts; study of structure or form.

mucilaginous (adj.) Of, relating to, full of, or secreting mucilage, a gelatinous substance that contains protein and polysaccharides.

multicellular (adj.) Having more than one cell.

multinucleate (adj.) Having more than one nucleus.

mycelium (pl. mycelia) Mass of hyphae that constitutes the body of a fungus or a funguslike protoctist.

mycology Study of fungi.

mycorrhiza (pl. mycorrhizae) Symbiotrophic physical connection between the hyphae of a fungus and the roots of a plant.

myxospore Spore or other desiccation-resistant stage of a myxobacterium.

nanometer (abbr. nm) 10^{-9} meter.

necrotrophy Nutritional mode in which a symbiotroph damages or kills the organism in which it resides.

nematocyst Cell of animals (e.g., ctenophores, cnidarians) containing a threadlike stinger—and in some cases, poisonous or paralyzing substances—used for anchoring, defense, or capturing prey.

nephridium (pl. nephridia) Excretory organ of many aquatic animals.

niche Role performed by members of a species in a biological community.

nomenclature System of identification and naming.

notochord Long elastic rod that is the internal skeleton in chordate embryos and that is replaced by the vertebral column in most adult chordates.

nucleic acid Long-chain molecule composed of nucleotides. Found esp. in cell nuclei, these molecules (e.g., DNA and RNA) are the basis of heredity and protein synthesis.

nucleoid DNA-containing structure of prokaryotic cells, not bounded by a membrane.

nucleoplasm Fluid contents of the nucleus of a cell.

nucleotide Compound composed of a sugar, a base, and a phosphate group; building block of nucleic acids.

nucleus (pl. nuclei) Large membrane-bounded organelle that contains most of the genetic information of a cell in the form of DNA.

nutrition Sum of the processes by which an organism takes in and utilizes food and energy sources—particularly carbon, nitrogen, phosphorus, and sulfur.

obligate (adj.) Compulsory; mandatory.

octopod A cephalopod mollusc that has eight arms with sessile suckers.

omnivory Mode of heterotrophic nutrition by which an organism obtains its food from a wide variety of microbes, plants, and animals.

oocyst Desiccation-resistant, thick-walled structure in which sporozoans are transferred from the tissue of one animal to that of another.

oogonium (pl. oogonia) Unicellular female sex organ that contains one or several eggs; female gametangium.

order Taxonomic level below class and above family.

organ Differentiated structure consisting of cells and tissues that are coordinated to form a specific function in an organism.

organelle Literally, "little organ"; distinct intracellular structure composed of a complex of macromolecules and small molecules; e.g., nucleus, mitochondrion, or undulipodium.

organic (adj.) Of, relating to, or containing reduced (hydrogen-rich) carbon compounds.

osmolyte Substance that maintains the salt balance in a cell.

osmosis Movement of a solvent through a semipermeable membrane into a solution of higher solute concentration that tends to equalize the concentrations of solute on the two sides of the membrane.

osmotrophy Mode of nutrition in which nutrients are obtained by absorption, the direct uptake of food molecules across membranes.

ostium (pl. ostia) Mouthlike opening into a bodily organ.

ovary Multicellular female reproductive organ that surrounds the egg(s).

ovule Structure in seed plants that contains the egg cell and, after fertilization, develops into a seed.

oxic (adj.) Containing molecular oxygen (referring to habitats).

oxidation Combination of a molecule with gaseous or atomic oxygen or the removal of hydrogen from a molecule.

oxygenic (adj.) Producing oxygen.

Paleozoic era Geological period from about 580 million to 245 million years ago.

papilla (pl. papillae) Small bump or projection.

parapodium (pl. parapodia) Fleshy, segmented appendage of polychaete annelids.

parasexuality Process that forms an offspring cell from more than a single parent without standard meiosis or fertilization.

parasite Organism that lives on or in an organism of a different species and obtains nutrients from it. Ill-advised term usu. referring to a symbiotroph with necrotrophic tendencies.

pathogen Organism that causes disease, usu. a microorganism that lives necrotrophically on or in an organism of a different species.

pellicle Thin, typically proteinaceous outer layer of a cell or organism, outside the plasma membrane.

pennate (adj.) Resembling a feather, esp. in having similar parts arranged on opposite sides of an axis like the barbs on the rachis of a feather.

peptidoglycan Rigid layer of bacterial cell walls.

perfect (adj.) Having both male and female reproductive organs on the same flower (referring to plants).

periplasm Peripheral cytoplasm. In prokaryotes, the space between the inner plasma membrane and the peptidoglycan layer of the cell wall.

periplastidial compartment Space between the plastid membrane and the plastid endoplasmic reticulum.

pH Scale for measuring the acidity of aqueous solutions; pure water has a pH of 7 (neutral); solutions with a pH greater than 7 are alkaline, less than 7 are acidic.

phaeodium (pl. phaeodia) Pigmented mass consisting primarily of waste products around the astropyle of the central capsule of phaeodarian actinopods.

phagocytosis Ingestion, by a cell, of solid particles by flowing over and engulfing them whole.

phagotrophy Mode of nutrition involving cell motility; active ingestion of particles via formation of food vacuoles.

pharynx (pl. pharynges or pharynxes) Throat; part of the digestive tract between the mouth cavity and the esophagus.

phloem Vascular tissue of plants that transports photosynthate, sugar, and other products of the leaves throughout the plant

phospholipid Lipid that contains phosphate esters. Cell membranes are made of layers of phospholipids in which proteins are embedded.

photic zone Region of surface waters in which enough sunlight penetrates to support photosynthesis. May extend 200 meters (approximately 600 feet) deep in clear lakes or open ocean or may be less than a centimeter in turbid water.

photoautotrophy Mode of nutrition in which all nutrient and energy requirements are met by using inorganic compounds and visible light; characteristic of cyanobacteria, algae, and plants.

photoheterotrophy Mode of bacterial nutrition in which all nutrient and energy requirements are met by using organic compounds and visible light.

photoplankton Free-floating microscopic or small aquatic organisms that are capable of photosynthesis; motile algae. (Same as phytoplankton, but better term because phototrophic bacteria and algae are not plants.)

photosynthate Chemical product of photosynthesis (e.g., glucose, starch).

photosynthesis Production of organic compounds from carbon dioxide and a hydrogen donor (e.g., water, hydrogen sulfide [H_2S], or hydrogen) using light energy captured by chlorophyll and relasing oxygen gas (O_2), sulfur globules, etc., as waste.

phototrophy Mode of nutrition in which light is the energy source.

phycobilin Water-soluble, protein-bound pigment, generally bluish or red, that is found in red algal plastids and cyanobacteria.

phycobiliprotein Complex of phycobilin pigments with protein.

phycobilisome Intracellular structure that contains phycobilin pigments and is arranged as a protrusion on the surface of or within the thylakoid membrane of a plastid.

phylogeny Evolution of a genetically related group of organisms; schematic diagram representing that evolution.

phylum (pl. phyla) Taxonomic level below kingdom and above class.

phytoplankton See *photoplankton*.

pigment Chemical substance made by the cell(s) or tissue of an organism body that imparts color.

pit connections Protoplasmic connections that join cells into tissues.

plankton Free-floating, microscopic or small, aquatic organisms; ecological, not taxonomic, term for some bacteria, protoctists, and animals.

planozygote Motile zygote of dinomastigotes; enlarged, undulipodiated, and sometimes thick-walled mastigote formed just after fusion.

plasma membrane Outer or cell membrane, composed of lipids and proteins, that surrounds a cell and regulates the exchange of material between the cell and its environment.

plasmodesma (pl. plasmodesmata or plasmodesmas) Minute cytoplasmic threads that extend through openings in cell walls and connect the protoplasts of adjacent living cells.

plasmodial reticulum (pl. plasmodial reticula) Cytoplasmic network through which ameboid cells move; characteristic of chlorarachnids.

plasmodium (pl. plasmodia) Multinucleate mass of cytoplasm lacking internal cell membranes or walls.

plastid Cytoplasmic, photosynthetic pigmented organelle (such as a chloroplast) or its nonphotosynthetic derivative (such as a leucoplast or etioplast).

polar capsule (1) In myxozoans, apical, thick-walled vesicle of a spore containing a spirally coiled, extrusible polar filament. (2) In heliozoan actinopods, region of dense cytoplasm that appears at opposite sides of the nucleus during mitosis.

polar filament Closed, tubelike structure coiled within the polar capsule of a myxozoan, which, when everted, has a sticky surface and may anchor the hatching spore to the surface of the intestine of the animal in which it resides.

polar membrane See *polar organelle.*

polar organelle Proteinaceous structure inside a bacterial cell (e.g., of *Cristispira* or *Desulfovibrio*) near the flagellum that is believed to be associated with motility.

polar tube Tubular, extruding organelle of microsporan spores that forcibly injects sporoplasm into single cells of animal tissue.

pollen Propagule; microspore of seed plants that contains a mature or immature microgametophyte (male gametophyte).

pollination Transfer of pollen from the stamen (the male reproductive organ) to the stigma (the female reproductive organ) in angiosperms.

polymer Chemical compound that consists of repeating structural units called monomers.

polymerase Enzyme that catalyzes the formation of polymers (e.g., DNA or RNA) by linking monomers.

polyp Life-cycle stage of cnidarians; cylindrical tube closed and attached at one end and open at the other end by a central mouth usu. surrounded by tentacles.

polypodial (adj.) Having more than one pseudopod.

polysaccharide Carbohydrate composed of many monosaccharide units joined to form long chains; e.g., starch and cellulose.

population A group of organisms of the same species in the same place at the same time.

predation Mode of heterotrophic nutrition that is necrotrophic but not symbiotrophic involving preying on moving bacteria, protoctists, or animals.

primary producer Ecological term for photo- or chemoautotrophic organisms; autotroph.

proboscis Tubular protrusion or elongation of the head or snout.

process Extension of a cell—e.g., a spicule, a heliozoan spine (axopod), or a foraminiferan reticulopod.

prokaryote Organism composed of cells that lack a membrane-bounded nucleus, membrane-bounded organelles, and DNA coated with histone proteins; member of the kingdom Monera (Prokaryotae).

propagule Unicellular or multicellular structure produced by an organism and capable of survival, dissemination, and further growth (e.g., cyst, spore, seed, and some types of egg).

protein Macromolecule composed of linked amino acids that is an essential constituent of all living cells.

Proterozoic eon Geological period from 2,500 million to 570 million years ago, preceded by the Archean eon and followed by the Phanerozoic eon.

protist Single-celled (or very-few-celled), and therefore microscopic, protoctist.

protoctist Nucleated organism that contains more than a single bacterially derived genome per cell but is not animal, plant, or fungus. Protoctists include the group of organisms traditionally called "protozoans" and all fungi with mastigote stages, as well as all algae (including kelps), slime molds, slime nets, and many other obscure eukaryotes. All are products of coevolved bacterial symbioses, and some, such as kelp, are too large to be called protists.

protoplasm Fluid contents of a cell (i.e., cytoplasm and nucleoplasm).

protoplast Nucleus and cytoplasm of a cell from which the cell wall has been removed.

protozoan Informal name of a member of the animal phylum Protozoa in the traditional two-kingdom (plant/animal) classification system consisting primarily of heterotrophic, microscopic eukaryotes (i.e., the smaller heterotrophic protoctists and their immediate photosynthetic relatives). (Obsolete term.)

pseudocilium (pl. pseudocilia) Protoplasmic protrusion of a cell that contains microtubules and is derived from the typical axoneme but is immotile; mastigoneme found on glaucocystophytes.

pseudoplasmodium (pl. pseudoplasmodia) (1) Structure resembling a multinucleate plasmodium that has retained its cell membrane boundaries. (2) Aggregate of amebas. (3) Uninucleate trophozoite cell containing one to several generative cells.

pseudopod Temporary cytoplasmic protrusion of an ameboid cell used for locomotion or phagocytotic feeding. Pseudopodium.

pseudopodium (pl. pseudopodia) See *pseudopod*.

pycnidium (pl. pycnidia) Asexual, hollow, dark-staining, multicellular structure lined on the inside with spore-bearing conidiophores; characteristic of fungi.

pyrenoid Proteinaceous organelle inside some plastids that is a center of starch formation.

radula (pl. radulae or radulas) Horny, toothed organ of molluscs that is used to rasp food and carry it into the mouth.

recombinant Organism or molecule derived from sexual processes such as fertilization or recombination of DNA.

red tide Sea water discolored by the presence of large growing populations of dinomastigotes. Blooms of some chrysophytes, euglenids, and the ciliate *Mesodinium rubrum* have also been correlated with red tides.

replication Process that increases the number of DNA or RNA molecules.

reproduction Process that increases the number of individuals. Asexual reproduction requires only one parent; sexual reproduction requires at least two parents.

respiration Oxidative breakdown of food molecules and release of energy from them in which the terminal electron acceptor is an inorganic compound such as oxygen or nitrate.

resting cyst Dormant life-cycle stage.

reticulopodium (pl. reticulopodia) Very slender, anastomosing pseudopod that is part of a network of cross-connected pseudopods; characteristic of foraminiferans.

rhizoid Rootlike structure in chytridiomycotes, algae, and many other protoctists that anchors and absorbs.

rhizome Subterranean or creeping plant stem that sends out shoots from near its tip or at nodes along its length.

rhodoplast Red plastid; the membrane-bounded photosynthetic structure of red algal cells.

rhyniophytes Group of fossil plants from the Devonian period that contains genera of the earliest vascular plants.

ribosome Spherical organelle composed of protein and ribonucleic acid; site of protein synthesis.

RNA Ribonucleic acid; a molecule composed of a linear sequence of nucleotides that can store genetic information; it is a component of ribosomes and plays a part in protein synthesis.

rumen (pl. rumina or rumens) One of the four chambers of the stomach of a cud-chewing mammal (ruminant) in which storage and initial digestion of food occurs.

sagenogen See *bothrosome*.

sclerotium (pl. sclerotia) Mass of hyphae, usu. with a darkened rind, that is capable of surviving unfavorable environmental conditions; characteristic of fungi.

sediment (1) Matter that settles to the bottom of a liquid. (2) Material that is deposited by wind, water, or glaciers.

seed Propagule; ripened plant ovule that contains an embryo.

septate junction Specialized area of adjoining cell membranes that show partitions in animal tissue.

septum (pl. septa) Dividing wall or membrane.

serial endosymbiosis theory (SET) Theory that all eukaryotic cells evolved from mergers among prokaryotes. According to the theory, undulipodia, mitochondria, and plastids began as swimming, respiring, and photosynthetic free-living bacteria that, in this order, established symbioses with *Thermoplasma*-like, fermenting bacteria.

sessile (adj.) Attached; not free to move about.

seta (pl. setae) Bristle.

sex Formation of a new organism from more than a single genetic source, usu. from the genetic endowment of two parents.

soredium (pl. soredia) Propagule; asexual reproductive structure of lichens; fragment containing fungal hyphae and algal cells.

sorocarp Multicellular, aerial, stalked structure derived from the aggregation of many individual cells; characteristic of slime molds.

sorus (pl. sori) Cluster of spores, sporangia, or similar structures in which spores are formed.

species Taxonomic level below genus that contains organisms that resemble each other greatly. In zoology, beings that can mate and reproduce with each other are placed in the same species.

sperm Male gamete, which is motile and usu. smaller than the female gamete.

spermatophore Capsule or packet that contains sperm.

spermatozoan Sperm; motile, undulipodiated male gamete of animals and some plants and protoctists.

spicule Small spine; slender, typically needle-shaped cell process.

spirillum (pl. spirilla) Flagellated bacterium with a helical form in which the flagella are external to the membrane.

sporangiophore Propagule-bearing structure; branch bearing one or more sporangia.

sporangium (pl. sporangia) Propagule-bearing structure; hollow, unicellular or multicellular structure in which propagules (cysts or spores) are produced and from which they are released.

spore Small or microscopic propagule, often but not always spherical, dark, desiccation- and heat-resistant, that contains at least one genome and is capable of further growth and eventual maturation.

spore stalk Propagule-bearing, multicellular, aerial, stalked structure derived from the aggregation of many individual cells; sporocarp; sorocarp.

sporocarp Propagule-bearing structure, usu. stalked, spore-bearing structure in which one initial cell is the source of all the spores.

sporogony Multiple-fission process of protoctists, esp. apicomplexans; multiple mitoses of a spore or zygote with no increase in cell size, producing sporozoites.

sporophore Propagule-bearing structure; protoctist stalk that bears spores.

sporophyll Plant leaf that bears sporangia.

sporophyte Diploid plant in which some cells undergo meiosis to form haploid spores.

sporoplasm Propagule of symbiotrophic protists; infective body. Ameboid organism within a spore.

sporozoan Informal name of a member of the animal class Sporozoa (in phylum Protozoa) in the traditional two-kingdom classification system; ambiguous former name for apicomplexans, which included the spore-forming parasites: myxozoans and microsporans. (Obsolete term.)

sporozoite Trophic propagule; motile, usu. infective product of multiple mitoses (sporogony) of a zygote or spore; characteristic of apicomplexans.

starch Complex, insoluble polysaccharide carbohydrate of some algae and green plants that constitutes one of their main energy reserve materials.

statoblast Propagule; asexually formed resistant internal bud of an ectoproct that, under severe conditions, is disseminated and grows into a new individual.

steroids Saturated hydrocarbons that contain seventeen carbon atoms arranged in a system of four fused rings. The hormones of the gonads and adrenal cortex, the bile acids, vitamin D, digitalis, and certain carcinogens are steroids.

stigma (pl. stigmata or stigmas) (1) In plants, the female flower part that is the receptive surface upon which pollen germinates. (2) In protoctists, an eyespot.

stolon Horizontal stalk near the base of a plant or colonial animal that produces new individuals by budding.

striated (adj.) Striped; referring to muscle marked by light and dark bands and made up of elongated, multinucleate fibers.

strobilus (pl. strobili) Modified ovule-bearing leaves or scales grouped together on an axis. Cone.

stromatolite Laminated carbonate or silicate rocks, organosedimentary structures produced by growth, metabolism, trapping, binding, and/or precipitating of sediment by communities of microorganisms, principally microbial-mat–forming cyanobacteria.

style Rigid, elongated organ or appendage (e.g., crystalline style).

stylet Small style.

substrate (1) Foundation to which an organism is attached (e.g., a rock). (2) Compound acted upon by an enzyme.

sulfuretum Marine habitat in which the sand smells of rotten eggs because of the bacterial reduction of sulfate to hydrogen sulfide.

suspension feeding Mode of heterotrophic nutrition; process of eating by filtering particles suspended in water.

swarmer cell See *zoospore*.

symbiogenesis The appearance in evolution of new forms of life resulting from symbioses, such as lichens from associations between algae and fungi. The fusion of lines on a phylogeny that represents new forms of life.

symbiont Member of a symbiosis.

symbiosis Intimate and protracted physical association between two or more organisms of different species.

symbiotrophy Mode of nutrition in which a heterotrophic symbiont derives both its carbon and its energy from a living partner.

synangium (pl. synangia) Organ; two or three sporangia joined together.

taxonomy Science of identifying, naming, and classifying organisms.

tentacle Elongate, flexible, usu. tactile appendage.

test Shell, hard covering, valve, or theca.

thallus (pl. thalli or thalluses) Simple, flat, leaflike body undifferentiated into organs such as leaves or roots.

theca (pl. thecae) Test; shell; hard covering.

thylakoid membrane Photosynthetic membrane, lamella, or sac that bears chlorophylls, carotenoids, and their associated proteins usu. stacked in layers.

tissue Aggregation of similar cells organized into a structural and functional unit; component of organs.

tornaria Larva; immature, ciliated form of certain hemichordates.

tracheophyte Vascular plant; plant that contains xylem and phloem.

trichocyst Organelle underlying the surface of many ciliates and some mastigotes that is capable of sudden discharge to sting prey.

trochophore Free-swimming, ciliated, marine larva.

trophic (adj.) Of or relating to nutrition; referring to feeding form in life-history stage.

trophomere Reproductive structure of ellobiopsids that bears gonomeres, which contain spores.

trophosome Digestive organ of submarine tube worms; midgut of vestimentiferans packed with symbiotrophic sulfide-oxidizing bacteria.

trophozoite Growing, feeding life-history stage of symbiotrophic protists.

tumor Amorphous swelling caused by uncontrolled growth of tissue cells with no apparent physiological function.

tun Cryptobiotic form of tardigrades that is shaped like a wine cask.

ultrastructure Detailed structure of cells and organs that is visible by transmission electron microscopy.

undulipodium (pl. undulipodia) Cilium or eukaryotic "flagellum"; the long, fine, intrinsically motile, intracellular structure of eukaryotes that is used for locomotion or for feeding. Undulipodia are underlain by [9(3)+0] microtubular kinetosomes from which they grow, as well as shaft microtubules in the [9(2)+2] array, and are composed of tubulin and other proteins (but not flagellin).

unicellular (adj.) Having or consisting of only one cell.

uninucleate (adj.) Having only one nucleus.

vacuole Space or cavity surrounded by a membrane in cell cytoplasm that contains fluid or air.

valve Test, shell, or hard covering.

vascular (adj.) Of or relating to conductive tissue.

vasoconstrictor Chemical or pharmaceutical (e.g., ergot) that reduces or constricts vertebrate blood vessels.

vegetative (adj.) (1) Resulting from mitotic cell divisions. (2) Of or relating to a growing and feeding stage of an organism's life history.

ventral (adj.) Of or relating to the anterior or lower surface (belly side) of an animal body.

vermiform (adj.) Resembling a worm in shape.

vertebra (pl. vertebrae or vertebras) One of the bony or cartilaginous segments that make up the spinal column.

vertebrate (adj.) Having a spinal column.

vibrio Bacterium shaped like a comma.

virus Protein-coated genetic material that is capable of growth and replication only within a living cell.

xanthin Yellow or orange carotenoid that is soluble in alcohol.

xanthophyll One of several yellow or orange oxygenated carotenoid pigments.

xanthoplast Plastid that contains xanthophylls.

xenoma Symbiotic aggregate formed by multiplying intracellular symbiotrophs within growing tissue cells, the whole structure increasing in size, as in the single-celled tumors formed by microsporans.

xylem Vascular tissue of plants that transports water and minerals from the roots to other parts of the plant; constitutes the wood of trees and shrubs.

zooecium (pl. zooecia) Shelter made of calcium carbonate skeleton overlaid with chitin and within which ectoprocts are anchored.

zoospore Undulipodiated, motile cell capable of germinating into a different developmental stage without being fertilized.

zygosporangium (pl. zygosporangia) Sporangium in which zygospores are produced.

zygospore Large, multinucleate, resistant structure (resting spore) that results from the fusion of two gametangia.

zygote Diploid nucleus or cell produced by the fusion of two haploid cells and destined to develop into a new organism.

Souces Used to Prepare the Glossary:

Abercrombie, M., M. Hickman, M. L. Johnson, and M. Thain. 1980. *The New Penguin Dictionary of Biology.* Eighth edition. Penguin, London.

Agrios, G. N. 1988. *Plant Pathology.* Third edition. Academic Press, San Diego.

Barnes, R. D. 1987. *Invertebrate Zoology.* Fifth edition. Saunders, Philadelphia.

King, R. C. 1990. *A Dictionary of Genetics.* Fourth edition. Oxford University Press, New York.

Lawrence, E. 1989. *Henderson's Dictionary of Biological Terms.* Tenth edition. Longman Singapore, Singapore.

Margulis, L. 1992. *Diversity of Life: The Five Kingdoms.* Enslow, Hillside, NJ.

Margulis, L., H. I. McKhann, and L. Olendzenski. 1993. *Illustrated Glossary of Protoctista.* Jones and Bartlett, Boston.

Margulis, L., and L. Olendzenski. 1992. *Environmental Evolution: Effects of the Origin and Evolution of Life on Planet Earth.* MIT Press, Cambridge, MA.

Margulis, L., and K. V. Schwartz. 1988. *Five Kingdoms: An Illustrated Guide to the Phyla of Life on Earth.* Second edition. W. H. Freeman, New York.

McFarland, W. N., F. H. Pough, T. J. Cade, and J. B. Heiser. 1979. *Vertebrate Life.* Macmillan, New York.

Pechenik, J. A. 1991. *Biology of the Invertebrates.* Second edition. Wm. C. Brown, Dubuque, IA.

Raven, P. H., R. F. Evert, and S. E. Eichhorn. 1986. *Biology of Plants.* Fourth edition. Worth, New York.

Webster's Ninth New Collegiate Dictionary. 1991. Merriam-Webster, Springfield, MA.

Webster's Third New International Dictionary of the English Language Unabridged. 1986. Merriam-Webster, Springfield, MA.

Bibliography

We have included here references to the biology literature on the diversity of life at levels from high school to professional and from old classics to works in preparation. Limitations of space, however, make impossible an adequate sampling of the enormous richness of published materials about life on Earth. In addition, much of the literature on natural history and science is published in other languages, whereas all the sources on this list are in English.

Listed alphabetically by author, these works have extensive bibliographies of their own, enabling the curious reader to delve even deeper into the literature. Here we list works by the following categories: (1) books, (2) articles, (3) classroom and laboratory teaching materials, (4) audiovisual materials, and (5) citation services. We describe briefly the contents of each entry; if illustrations are not mentioned, assume that the work has few or none. Each work is also identified by one of the following codes: AT (adult trade), ATX (adult text or reference, nonspecialized), ATX-S (adult text or reference limited to readers with considerable previous experience in the subject matter), and YA (young adult readers).

Information about obtaining specific material is available from the publisher or the distributor. Librarians and bookshop keepers most often can help you find what you want. Some of the most useful tools that they use include a reference set called *Books in Print,* the biological or chemical abstracting services, and ISI (the Institute for Scientific Information, see page 215), which maintains on computer a huge data base of scientific publications that is available in many forms (e.g., print, floppy disk, magnetic tape, or compact disk).

Books

Balows, A., H. G. Trüper, M. Dworkin, W. Harder, and K.-H. Schleifer, eds. 1991. *The Prokaryotes. A Handbook on the Biology of Bacteria: Ecophysiology, Isolation, Identification, Applications.* 4 volumes. Second edition. Springer-Verlag, New York. This exhaustive, illustrated listing of bacterial groups is in large format with professional references. ATX-S.

Barlow, C., ed. 1991. *From Gaia to Selfish Genes: Selected Writings in the Life Sciences.* MIT Press, Cambridge, MA. Sparsely but artistically illustrated, this book gives a wonderful flavor of the nature of the debate among life scientists and others about what life is and how it works. Insightful commentaries by the editor establish a context for a wide range of well-stated opinion. AT.

Barnes, R. D. 1987. *Invertebrate Zoology.* Fifth edition. Saunders, Philadelphia. This standard college text systematically reviews and illustrates all phyla of aquatic animals and nonvertebrate chordates. ATX.

Cloud, P. 1988. *Oasis in Space: Earth History from the Beginning.* W. W. Norton, New York. Profusely illustrated with maps, charts, diagrams, and photographs, this book is the master work of a great geologist who understood the first 80 percent of Earth's record, the pre-Phanerozoic time division during which life evolved and developed most of its tricks of planetary modification. Through this tour of the Hadean, Archean, and Proterozoic eons, we realize that the beginning of the story of biodiversity and environment is far too interesting to miss. ATX.

Francki, R. I. B., C. M. Fauquet, D. L. Knudson, and F. Brown, eds. 1991. *Classification and Nomenclature of Viruses: Fifth Report of the International Committee on Taxonomy of Viruses.* *Archives of Virology,* second supplement. Springer-Verlag, Vienna. Written by teams of professional virologists, this all-inclusive tome gives names of viruses and criteria for grouping them. ATX-S.

Groombridge, B., ed. 1992. *Global Biodiversity: Status of the Earth's Living Resources.* A report compiled by the World Conservation Monitoring Centre. Chapman & Hall, London. This large book, which contains colored biodiversity maps, is a report developed in collaboration with the Natural History Museum, London; the World Conservation Union; the United Nations Environment Program; the World Wild Fund for Nature; and the World Resources Institute. With no index and only a very limited glossary, it is still a most up-to-date tabulation of the flora and fauna of the world. Although not entirely absent, the discussion of the microbiota is weak. The book is, however, extremely useful for an international perspective on biodiversity. ATX-S.

Joseph, L. E. 1990. *Gaia: The Growth of an Idea.* St. Martin's Press, New York. This story of how the idea of Gaia originated and was contested and co-opted by nonscientists gives the reader great insight into the relationship between specialist science and the ideas of the society that supports it. AT.

Kendrick, B. 1992. *The Fifth Kingdom.* Second edition. Mycologue Publications, Waterloo, Ontario. Illustrated with simple diagrams and black-and-white drawings, this book is a modern description of fungal diversity and structure in taxonomic order. ATX.

Krieg, N. R., and J. G. Holt. 1984. *Bergey's Manual of Systematic Bacteriology,* volume 1. Williams & Wilkins, Baltimore. Nicknamed the microbiologist's "Bible," this standard reference has been expanded in recent years to include cyanobacteria and archaebacteria—in fact all formerly recognized bacteria—and now occupies several volumes. Reflecting the point of view of the professional microbiologist, the work's main purpose is to identify bacteria important to medicine, food processing, soil science, and other practices. ATX-S.

Lederberg, J., ed. 1992. *Encyclopedia of Microbiology.* Academic Press, San Diego. A four-volume, multi-authored work on topics ranging from AIDS in Africa to alkaliphilic bacteria, refrigerated foods, and wine, this text is less an encyclopedia than it is a smorgasbord of essays that give the reader a good idea of the state of microbiology today. ATX-S.

Lee, J. J., S. H. Hutner, and E. C. Bovee. 1985. *An Illustrated Guide to the Protozoa.* Society of Protozoologists, Lawrence, KS. Profusely illustrated primarily with drawings and scanning electron micrographs, this book is particularly good for the lower taxa of ciliates, dinomastigotes, and other small heterotrophic protoctists. The protozoa are classified here in the animal kingdom, so the occasional inclusion of algae, slime molds, and other protoctists that traditionally have been classified as plants or fungi is inadequate, as their treatment is prejudicially zoological. ATX-S.

Lovelock, J. E. 1988. *The Ages of Gaia: A Biography of Our Living Earth.* W. W. Norton, New York. The most articulate and riveting description of the idea that life, a planetary-level phenomenon, creates and alters its own habitat was written by the inventor of the idea that Earth is alive. Because Lovelock traces the planetary-level influence of life and its history both from its beginning and from his beginning as a student of the process, the book speaks in a personal way to the reader. AT.

Lowenstam, H. A., and S. Weiner. 1989. *On Biomineralization.* Oxford University Press, New York. Over 50 minerals are produced by organisms at temperatures and pressures that preclude their formation geologically. Details of the distribution, formation, structure, geological importance, and other aspects of these organisms are found in this unique advanced work. ATX-S.

Margulis, L. 1992. *Diversity of Life: The Five Kingdoms.* Enslow, Hillside, NJ. With many black-and-white illustrations, this small book details the major features, members, and history of the five kingdoms and provides the background for the five-kingdoms poster classroom activities (see page 214). YA.

Margulis, L., J. O. Corliss, M. Melkonian, and D. J. Chapman. 1990. *Handbook of Protoctista: The Structure, Cultivation, Habitats and Life Histories of the Eukaryotic Microorganisms and Their Descendants Exclusive of Animals, Plants and Fungi.* Jones and Bartlett, Boston. The "Bergey's Manual" (see Krieg and Holt, 1984) of the eukaryotic microorganisms, this large-format, profusely illustrated book, the work of 50 authors, integrates information on acquisition, culture, structure and classification, the fossil record, the literature, and other aspects of these organisms. ATX-S.

Margulis, L., and K. V. Schwartz. 1988. *Five Kingdoms: An Illustrated Guide to the Phyla of Life on Earth.* Second edition. W. H. Freeman, New York. One page of labeled illustration analyzes the structure of a typical member of each phylum of bacteria, protoctists, fungi, animals, and plants with a description of major features. In this book, which succinctly describes the highest taxa, all phyla are represented, nearly 100 inclusive groups of organisms. New discoveries about archaebacteria are integrated into the rest of biology. ATX.

McFarland, W. N., F. H. Pough, T. J. Cade, and J. B. Heiser. 1979. *Vertebrate Life*. Macmillan, New York. Organized systematically—that is, based on evolutionary patterns—this illustrated college text describes all the vertebrate members of the chordate phylum, animals of the most interest to people. ATX.

National Research Council. Committee on Research Opportunities in Biology. 1989. *Opportunities in Biology*. National Academy Press, Washington, DC. About 450 well-indexed pages explain the new, chemically based biology, cell and development study, medically oriented biology, evolution and diversity, ecology and ecosystems, plant biology, and agriculture as the professional views these subfields. This book is ideal for anyone studying biology, no matter at what level, especially for those who wish to enter the field. YA.

Pechenik, J. A. 1991. *Biology of the Invertebrates*. Second edition. Wm. C. Brown, Dubuque, IA. This college text of zoology depicts all major groups of aquatic and terrestrial animals, exclusive of the vertebrates, in what is thought to be their evolutionary context. ATX.

Raven, P. H., R. F. Evert, and S. E. Eichhorn. 1992. *Biology of Plants*. Fifth edition. Worth, New York. Beautifully written and illustrated, this description of all major groups of plants (and protoctists traditionally classified with plants), their structures, life cycles, and likely evolutionary histories, is especially suited for college students seeking botanical knowledge. ATX.

Sagan, D., and L. Margulis. 1993. *Garden of Microbial Delights: A Practical Guide to the Subvisible World*. Second edition. Kendall-Hunt, Dubuque, IA. Abundantly illustrated, this book is a guide to the history of knowledge, diversity, and usefulness of the microscopic world, including notes on how to keep microbial pets. YA.

Schneider, S. H., and P. J. Boston, eds. 1991. *Scientists on Gaia*. MIT Press, Cambridge, MA. Although published a few years later, these far-reaching papers (with only a few specialized illustrations) are the result of an exciting and well-attended Chapman Conference in 1988, the first time ever that the Gaia idea and its ramifications were discussed openly by professional scientists. The diversity of coverage and opposition of philosophies give the reader an idea of how geophysicists, climatologists, ecologists, and other professionals tend to regard Earth as an object of their study. ATX-S.

Sneath, P. H., N. S. Mair, M. E. Sharpe, and J. G. Holt. 1986. *Bergey's Manual of Systematic Bacteriology*, volume 2. Williams & Wilkins, Baltimore. See Krieg and Holt (1984). Volumes 3 and 4 have now also appeared. ATX-S.

Sonea, S., and M. Panisset. 1983. *A New Bacteriology*. Jones and Bartlett, Boston. This small, fascinating book is a long essay about the dispersed nature of the global bacterial organism and its strange genetic system in an enormous diversity of habitats. ATX.

Westbroek, P. 1991. *Life as a Geological Force: Dynamics of the Earth*. W. W. Norton, New York. Small and charming, with enchanting pencil illustrations, this book reveals the environment through the eyes of a Dutch geologist familiar with the world and its record of life as registered by fossils, sediments, and ancient volcanoes. From plate tectonics to the Scottish highlands and carbonate slime in the Caribbean, Westbroek explores ideas of the biosphere, Gaia, and attempts of scientists to make models of the complexity of the real world. AT.

Wilson, E. O., ed. 1988. *Biodiversity*. National Academy Press, Washington, DC. Available in softcover, this book is the report of a symposium on the loss of diversity due primarily to the expansion of human ecosystems. Forests, coastal zones, "deep ecology," the monitoring of diversity and its value, attempts to restore lost ecosystems, and the varying views of nature are discussed in short papers by many well-qualified authors. A video of a portion of this symposium is available from the National Academy. ATX.

Wilson, E. O. 1992. *The Diversity of Life*. Belknap Press of Harvard University Press, Cambridge, MA. Perhaps the most moving and authoritative recent work on the value of nature in the splendor of its diversity, this book was written by the master of the subject. Charming, beautifully drawn, black-and-white illustrations help document the seriousness of species loss by habitat disturbance and destruction. AT.

Articles

Margulis, L. 1992. Biodiversity: Molecular biological domains, symbiosis and kingdom origins. *BioSystems* 27:39–51. This article lists technical definitions of the five kingdoms, their phyla, criteria for classification, and first appearance in the fossil record. ATX-S.

Wilson, E. O. 1993. Is humanity suicidal? *New York Times Magazine*, May 30. A moving, totally comprehensible description of the dilemma of the ecological calamity, this article is written by the author of two books listed above, a superb scientist who has spent his life in the field of the far-flung wildernesses, the laboratory, and the classroom. YA.

Classroom and Laboratory Teaching Materials

TEACHER'S GUIDES

Armstrong, L., and L. Margulis. 1992. *Teacher's Guide to the Five Kingdom Poster.* Ward's Natural Science Establishment, Rochester, NY. This guide describes classroom activities to accompany the *Five Kingdoms* poster (drawings by Christie Lyons based on design by Dorion Sagan). YA.

Margulis, L., et al. 1992. *What Happens to Trash and Garbage? An Introduction to the Carbon Cycle.* Ward's Natural Science Establishment, Rochester, NY, which includes Margulis, L., and L. Olendzenski. 1991. *Common Fungi: Teacher's Guide* [videocassette]. This boxed unit provides from one lesson to several weeks of activities that link our ordinary experience with waste materials to the movement of matter, especially inorganic and organic carbon, through the biosphere, atmosphere, and soil, focusing especially on the roles of fungi and people in the carbon recycling process. A poster of the carbon cycle, photographs to be matched with appropriate captions, and other materials for associated activities come with the unit. YA.

Margulis, L., and D. Sagan. 1988. *The Microcosmos Coloring Book.* Harcourt Brace Jovanovich, Boston. With labeled drawings and explanation in the text, this entry into the strange world of the subvisible illustrates microbes of the seashore, forest, karst, and park (i.e., those living on people). YA.

PROJECTION SLIDES

The following slide sets, in color, are accompanied by teacher's guides that include worksheets. Single sets are packed in translucent, plastic sheets; the entire set of six comes in a notebook that includes pockets for the written materials.

Margulis, L., and K. V. Schwartz. 1987. *Introduction to the Five Kingdoms* [20 35-mm slides and teacher's guide]. Ward's Natural Science Establishment, Rochester, NY. The history, distribution, and examples of the five kinds of life, as well as cell and virus structure, are presented.

Margulis, L., and K. V. Schwartz. 1987. *Monera* [40 35-mm slides]. Ward's Natural Science Establishment, Rochester, NY. The major types of bacteria, prokaryotic cell structure, and environmental effects of bacteria are explored.

Margulis, L., and K. V. Schwartz. 1988. *Protoctista* [40 35-mm slides]. Ward's Natural Science Establishment, Rochester, NY. Examples of each major type of protoctist—algae, slime molds, ciliates, amebas, and others—photographed live, are shown.

Margulis, L., and K. V. Schwartz. 1987. *Fungi* [20 35-mm slides]. Ward's Natural Science Establishment, Rochester, NY. From mating hyphae of basidiomycotes through the resulting mushrooms to lichens and their components, this slide set shows the major features of fungi, their life cycles and beauty.

Margulis, L., and K. V. Schwartz. 1988. *Animals* [40 35-mm slides]. Ward's Natural Science Establishment, Rochester, NY. An example of a member of each phylum of animals is shown alive in its habitat.

Margulis, L., and K. V. Schwartz. 1987. *Plants* [40 35-mm slides]. Ward's Natural Science Establishment, Rochester, NY. An example of a member of each phylum (division) of plants is shown. Plant habitats and structure are illustrated.

Audiovisual Materials

Margulis, L., and L. Olendzenski. 1991. *Common Fungi: Teacher's Guide* [videocassette]. Ward's Natural Science Establishment, Rochester, NY. The development of *Rhizopus* on bread, *Fusarium, Alternaria,* and *Coprinus* from spores is depicted. Hyphae, mycelia, growth rings, pigments, and other features of live fungi are illustrated. Each numbered scene via time-lapse photography is described in film notes. 16 minutes; silent.

Our Living Planet series [videocassette]. 1993. Ward's Natural Science Establishment, Rochester, NY. YA.

> Part I: *Origins of Life.* Music and graphics are included in this explanation of modern ideas of how life began and what life is. 10 minutes; color.

> Part II: *Five Kingdoms of Life.* With the help of live examples, models, and graphics and accompanied by music, this video reviews members of the five kingdoms and their interactions. 10 minutes; color.

> Part III: *People Are . . . Mammals.* On a journey beginning at the level of kingdom (the animals) and traveling through the chordates, vertebrates, tetrapods, mammals, and primates to *Homo sapiens,* this video explains why people are classified as they are. 10 minutes; color.

Citation Services

Current Contents: Life Sciences. Institute for Scientific Information, Philadelphia. Copies of the tables of contents of hundreds of scientific periodicals are published weekly.

Science Citation Index. Institute for Scientific Information, Philadelphia.

Acknowledgments

We are grateful to the hundreds of biologists whose research makes this compendium possible. Help with the manuscript came from Glyn Davies, Mark McMenamin, Lorraine Olendzenski, Donna Reppard, Dorion Sagan, Jonathan Schwartz, and Oona West. We are also grateful to Donna for drawing the Forest Clearing illustration (page 112) on short notice. To Lowell Schwartz and Sona Kim-Dolan, our thanks for urging us onward and for illuminating insights. Dr. Jorge and Sylvia Potter provided the enthusiasm of ardent naturalists. This work has been supported by the NASA Life Science Office, the Richard Lounsbery Foundation of New York City, and the University of Massachusetts Department of Biology in the College of Natural Science and Mathematics.

Index

Boldface numbers refer to illustrations.